The New Naturali

A SURVEY OF BRITISH NA

BRITISH BATS

The aim of this series is to interest the general reader in the wildlife of Britain by recapturing the enquiring spirit of the old naturalists. The editors believe that the natural pride of the British public in the native flora and fauna, to which must be added concern for their conservation, is best fostered by maintaining a high standard of accuracy combined with clarity of exposition in presenting the results of modern scientific research.

The New Naturalist

BRITISH BATS

John D. Altringham

With 8 colour plates and over 120 black
and white photographs and line drawings

HarperCollins*Publishers*

HarperCollins*Publishers*
77–85 Fulham Palace Road
Hammersmith
London W6 8JB

The HarperCollins website address is:
www.**fire**and**water**.com

Collins is a registered trademark of HarperCollins*Publishers* Ltd.

First published 2003

ISBN 000 220140 2 (Hardback)
ISBN 000 220147 X (Paperback)

Printed and bound in Great Britain by the Bath Press

Contents

Editors' Preface

Britain now has fewer than 50 species of truly native land mammals and of these more than a third are bats. Given the important place that they occupy in our fauna it might be thought surprising that the New Naturalist series has not produced a volume on bats long before this. In fact the editorial archive reveals that the subject of a bat volume was first discussed in 1959. Ten years later the idea was still on the table when it was concluded that there was insufficient demand to proceed with the project! The last two decades have witnessed a huge increase in interest in our bats fuelled by a combination of academic research and the activities of organisations such as the Bat Conservation Trust and local bat groups. So it is with an unusually high sense of satisfaction that the Editors welcome this addition to the New Naturalist library, some 43 years after it was first mooted and during which time four new species of bats have been added to the British fauna!

Bats have not always been held in a particularly high regard in the popular imagination, being associated with witches and vampires and the forces of darkness. However, this is but one of many aspects of bat lore where all is not what it seems. Most naturalists will know that our commonest bat by a long way is the pipistrelle. Yet even this apparently secure piece of natural history now has to be revised as modern techniques of ultrasound recording have revealed that where there was one there are now two species.

Bats are unique among mammals in the evolution of powered flight and this, together with their possession of a highly developed system of echolocation by which they navigate and capture their prey, serves to further increase their fascination. We were particularly fortunate to find the ideal author to bring to life and explain these special areas of bat biology and to bring us up to date with the most recent ideas. John Altringham is Professor of Biomechanics in the School of Biology at the University of Leeds. He is a distinguished authority on the mechanics of animal locomotion which he combines with a particular interest in the biology and natural history of bats. A particular feature of the book is his discussion of bat communities and the ways in which the different species interact. This approach is of critical importance for those concerned with habitat management for bats.

The final chapters of the book give a wealth of advice on how to develop and further an interest in these remarkable creatures. Detailed accounts of all of the British species are followed by hints on practical projects, equipment, conservation and identification, and the law. Throughout, the book is enriched by Tom McOwat's superb illustrations and Frank Greenaway's stunning photography.

Author's Foreword and Acknowledgements

I was surprised and delighted to be asked to write this book, being well aware of the long tradition behind the New Naturalist titles. Despite this long tradition, the New Naturalists is a very diverse series, so I did not feel bound by convention to a particular style or format. How then would I write this book? As a research scientist I am trained to ask questions and to attempt to explain what I see. I want to understand how and why the natural world works as it does, fitting my observations into a scientifically robust evolutionary framework. My aim has therefore been to describe the fascinating natural history of bats, but to do so in a functional and evolutionary context. This has meant introducing some concepts and processes that may not be familiar to all amateur natural historians. However, I believe this approach helps to bring meaning to natural history, making it more interesting and cohesive. Potentially unfamiliar concepts and terms are explained as they are introduced and I have also provided a glossary at the end of the book. My hope is that this investigative approach will be taken up by the more active bat naturalists. To this end I have made some suggestions for research that amateurs can tackle. I have also discussed equipment and techniques in some detail. However, first and foremost, this is a book about the natural history of British bats, a New Naturalist title that is long overdue, given the prominent place bats have in our native fauna. I only hope that my efforts have done the subject justice.

I knew from the start who would illustrate the book. It is a feast to the eye, thanks to Frank Greenaway's stunning photographs in the centre of the book and Tom McOwat's beautiful drawings. I will not forget the evening at Frank's, sitting back after a good dinner to the difficult task of choosing these few photographs from the hundreds on offer. I gave Tom an extravagant wish list of possible drawings, hoping he would find the time to tackle at least half of them. I should have known that he would do all of them, and all to his exacting standards.

I have worked with bats for almost 20 years with an ever-growing band of fellow enthusiasts. I would like to thank them for making it so much fun and for helping me acquire the knowledge I needed to write this book. I cannot name them all, but some deserve a special mention. The cast, in approximate order of appearance, is: David Bullock, Charles Critchley, Kirsty Park, Gareth Jones, Geoff Billington, Dean Waters, Ruth Warren, Brock Fenton, Chris Wright, Kathy Meakin, Paula Senior, Nicky Green, James Aegerter, Anita Glover and Eleni Papadatou.

Isobel Smales edited the book with a light but firm hand, drawing my attention to a number of clumsy errors, reining me in once or twice when I got carried away on peripheral issues and pointing out the need for further explanation where I had provided too little. Kirsty Park and Paula Senior read the entire manuscript and provided valuable comments. Thanks also to the many people, unfortunately too many to name, who gave me access to unwritten thoughts or unpublished information. Tony Mitchell-Jones provided valuable summaries of the English Nature hibernacula database.

Balancing family life with a day job in research and teaching, and a night job chasing bats, requires some effort. It is a measure of the patience and understanding of my wife Kate that I was able to add the additional burden of this book to my long list of commitments, knowing I would get encouragement and help as well as understanding.

John D. Altringham

1

From Dark Obscurity

There are over 4,000 species of mammal in the world, and almost one in four of them is a bat. In Britain there are fewer than 50 truly native 'land' mammals, and one in three of these is a bat. Two species are so common, widespread and easily seen that they must be our most readily observable native mammals. Few mammals live closer to humans than bats. Many species roost unnoticed in our homes, and some are now almost entirely dependent on built structures for their survival. Bats are unique among mammals in being capable of powered flight. They are also one of just two groups that have a sophisticated echolocation system (the other being the dolphins and their relatives). Few mammals are more accomplished hibernators. So, here is a group of animals that constitutes a major component of our mammalian fauna, has widespread and common members, lives close to humans, and is biologically unusual and fascinating.

Why then, until recently, have relatively few naturalists taken an interest in bats? Is it the antisocial hours they keep? Only the most dedicated naturalist works far into the night. But this cannot be the answer: most bats are active at dusk, and many of our earthbound mammals keep equally unsociable hours. Perhaps it is because we do not see them often enough – it is hard to get close to a bat. But it is hard to get a close look at most of our smaller mammals, whether nocturnal or diurnal, and all are more timid than bats. In fact, bats will continue to forage in the presence of large groups of noisy, torch-wielding naturalists, and you can sit and watch them for hours – this can be said of few other mammals. Is it because it is difficult to identify flying bats? It is true that a flying bat can rarely be identified to species, and typically only be put into a category with several others. But there are many species out there, some with a distinctive flight style, and, with the help of inexpensive bat detectors, some can be identified. Maybe it is because they are hard to study. I doubt this too, and I will argue later that the keen naturalist can learn more about bats through his or her own efforts than they can about almost any other mammal. It is therefore difficult to come up with a rational reason for the long neglect of bats. On the other hand, it is quite easy to explain their current popularity.

Knowledge and understanding may have been late in coming, but people have long been fascinated by bats. In western cultures this has been based largely on fear and superstition: Bram Stoker has a great deal to answer for! Before the publication of *Dracula* in 1897, bats were not linked with witches, vampires and the evil side of the supernatural in any significant way. In European traditions, vampires were everything but bats: horses, dogs, spiders, frogs, snakes, even sheep, to name but a few. In the same gypsy cultures bats were often seen as symbols of good luck. This did not protect them: various parts of bat were carried around as charms. In the rest of the world bat mythology is richer, and bats were as often seen as omens of good as of evil. India has one of the strongest vampire traditions, but bats are never mentioned. Only in

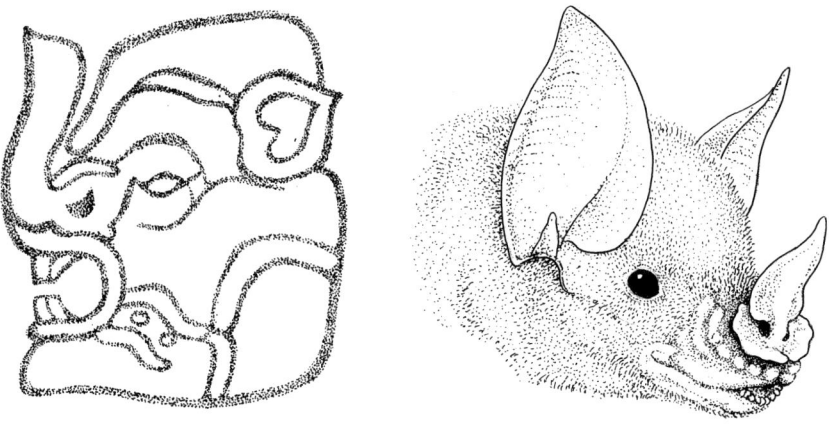

Fig. 1.1 Left: Zotz, the vampire bat god from a Mayan carving. Right: a typical phyllostomid, *Phyllostomus hastatus.*

the New World is there a tradition of bats as vampires, and this of course has a firm biological basis: only in South and Central America are real vampire bats found. In Mayan mythology, Zotz, the vampire bat god, lived in the underworld, through which humans passed as they died. Zotz shows a striking resemblance to the spear-nosed bats, a numerous and diverse family of neotropical bats (Phyllostomidae), of which the vampire is a member (Fig. 1.1).

To most cultures, the night is a time of mystery, fear and often death, and bats have inevitably been linked with these matters. To the ancient Greeks, bats were associated with Persephone, wife of Hades, ruler of the underworld. An Australian Aboriginal legend tells of how the first man and woman were warned not to disturb a large bat venerated by the spirits. Curiosity got the better of the woman, who frightened the bat from its perch in front of a cave. The bat was the guardian of death, which was able to escape from the cave, bringing mortality to humans. Another Aboriginal myth relates how a fight between the birds and the 'animals' caused the god Yhi to hide the light. The bat, who had been led astray by the treacherous owl, made amends by cutting the darkness in two with a boomerang. He gave light to the birds and animals and took the darkness for himself. The owl was never forgiven, and is to this day mobbed by the birds as he flies out at dusk. Myths in many cultures explain why bats fly at night, and why they appear to be neither bird nor mammal.

The traditional Chinese view of bats is the most positive. The bat is a symbol of good luck and happiness, and the common motive of five bats around a peach tree, the Wu-Fu (Fig. 1.2), represents the five great happinesses: health, wealth, good luck, long life and tranquillity.

Most of these myths are accepted for what they are, but some linger on as truths in the minds of many. It is still a widely held belief that bats are blind, that they get caught in your hair, or that vampire bats are everywhere, even in Britain. Thankfully, most people are now better informed.

Why have bats come out of their dark obscurity and into the light over the last two decades? Even now, initial curiosity is most often aroused by the myth and mystery that surround bats. However, more and more people are discov-

Fig. 1.2 Wu-Fu, Chinese symbol of the five great happinesses: health, wealth, good luck, long life and tranquillity.

ering that the truth about bats is far more interesting than the myth. The efforts of a dedicated group of professional and amateur biologists have played a major role in increasing public enlightenment. *The Wildlife and Countryside Act 1981* gave protection to all UK bats and their roosts, and was a catalyst to the growth of the county bat groups. Fauna and Flora International (the Fauna and Flora Preservation Society as it then was), the Mammal Society and the Vincent Wildlife Trust have played important roles in promoting bats, and continue to do so. The 1990s saw the formation of the Bat Conservation Trust out of the Bat Groups of Britain, to assist the work of the bat groups, and to carry out its own programme of activities.

Advances in the technology used in both research and television bring newly discovered and fascinating aspects of bat biology right into people's homes. A decade ago, bats made very rare appearances in natural history films. They are now a common feature, and several excellent programmes have been devoted to bats. New equipment accessible to the amateur as well as the professional brings us nearer to bats in the wild, increasing our understanding and hence our appreciation of these animals. Ultrasound or bat detectors enable us to listen to bats (p. 161). Even the most primitive of detectors alerts us to the presence of bats and can help to identify species. Equipment costing no more than a good pair of binoculars can be used to make high quality recordings of echolocation and social calls, opening up new avenues of research and understanding to the amateur. For a similar price, adequate night vision equipment is also now available, and digital camcorders now have 'nightshot' facilities.

All these advances have helped bring bats into the light, and for many people the transition from ignorance to enthusiasm is swift and enduring. My own story illustrates this well. Although a professional biologist and a keen naturalist, I knew little about bats. Then, in the summer of 1984, I was taken to a Natterer's bat roost. The trap was sprung instantly. As I examined the delicate but energetic animal in my hand it buzzed, struggled a little and then quietly dozed. I was hooked: I had to know more. As a researcher in biomechanics and animal locomotion, bat flight and echolocation had instant appeal, but I was equally enthralled by their enormous diversity and, to me, incomparable natural history. The more I read, the more I wanted to read. My interest in bats has retained an amateur flavour; a personal enthusiasm squeezed between my professional and personal lives, and part of both. What makes it worth the effort in an already hectic life is the quiet excitement of the field work and the shared enthusiasm.

Another factor is the constant awareness of just how little is known about our bats. Take the common pipistrelle. Over the last few years it has become apparent that it is two distinct species, with quite different lifestyles – the first new mammal species to be identified in the country for a very long time. Although we know where female pipistrelles set up their summer nursery roosts, we have little idea about where the males roost in summer. We have even less idea about where both sexes hibernate – our most common bats all but disappear in winter. The picture is similar for most species. Both amateur and professional naturalists have no shortage of questions to ask and to seek answers to.

The study of the natural history of bats is fulfilling in its own right, but there is a more practical side to it. There is good evidence, if often circumstantial, that many of our bat species are in decline. Action is needed if this decline is to be stopped. This action can take many forms, but fundamental to all our conservation efforts is the need for a deeper knowledge of the biology and natural history of our bats. A naturalist's enthusiasm comes from an understanding of his or her subject. To infect others with this enthusiasm, this knowledge needs to be passed on. It is a tool for informing, inspiring and persuading. Good conservation relies on this knowledge. The best conservationists, at all amateur and professional levels, know their biology and are good natural historians. Conservation organisations have been enormously effective in making the public aware that bats need our help, and have increased the public understanding of bats. However, I believe that in this and other areas of conservation, awareness and support have often preceded knowledge. Too many people know too little about what they agree needs conserving. Many people may know all they want to know, but there are others eager to learn. The more that is known about our wildlife and environment, the better we can conserve what we still have, and the more our own lives are enriched and enjoyed.

My aim is to provide a readable overview of the natural history of our native bats. If you are not already a bat naturalist, I hope it will persuade you to get out and see and hear bats for yourself. As small, nocturnal and very mobile mammals go, bats are surprisingly accessible and wonderfully fascinating. I believe the first step in nature conservation, and the best, is to grip your audience with the fascination of nature itself. The need for conservation is then self-evident, without the need for heavy-handed preaching or a homocentric appraisal of why wildlife is important to *us*. I hope that this book goes a little way towards fulfilling this first step.

2

Bats, an Evolutionary Success Story

British bats: their place in the bat world

Our bat fauna comprises just 16 of the world's 1,000 or so species. If we are to appreciate our own bats, we must view them as part of a large, widespread and diverse mammalian order. Bats are arguably the most successful and diverse mammals ever to evolve, and have much to teach us about the natural world. As with virtually all other animals and plants, the number of species is greatest in the tropics and declines as you move north or south to the poles. Most tropical regions have between 100 and 200 species, North America has a few over 40, and north-west Europe a few less. Britain's 16 species fit the pattern well.

Worldwide there are 18 families, of which two have representatives in Britain: vesper or evening bats (Vespertilionidae) and horseshoe bats (Rhinolophidae). All British bats are insectivores, as are almost 800 other species. However, those remaining feed on everything from fruit, flowers, nectar and pollen to fish, birds and blood. In our cold and inhospitable winters both vesper and horseshoe bats hibernate, but hibernation is rare among bats, and no members of the other 16 families are confirmed hibernators. Finally, all our bats are small, ranging from 4 to 30 grams (for example, pipistrelle to noctule): the smallest bat in the world is less than 2 grams, the largest over 1 kilogram. This chapter considers the origins and current diversity of the world's bats.

The origins of bats

The most direct evidence about the origins of an animal is to be found in fossils. By a remarkable coincidence, the best-preserved fossil bats are also the oldest. *Icaronycteris index* is a small insectivorous bat from the Eocene period, 50 million years ago (Fig. 2.1). It was beautifully preserved in the Polecat Bench rocks of Wyoming in the United States. Equally well preserved are specimens of seven insectivorous bats from Germany. They also date back to the Eocene, from 45 million year-old oil-shale pits at Messel near Darmstadt in Germany. The chemical conditions and fine-grained rock have retained stunning detail in the Messel fossils. When examined under the electron microscope, the insects that were a bat's last meal can often be seen, and, on the wingscales of the moths, pollen from the flower from which the moth perhaps had *its* last meal – ecology preserved in stone.

So what do these fossils tell us about the origins of bats? Most obviously that bats 50 million years ago were much like modern bats: the Eocene bats were clearly advanced. Although they show some primitive features, they have almost all the distinguishing features of modern bats. They were obviously strong and agile flyers, and had well-developed echolocation (revealed by X-ray studies of their hearing system). This leaves three important questions

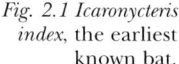

Fig. 2.1 Icaronycteris index, the earliest known bat.

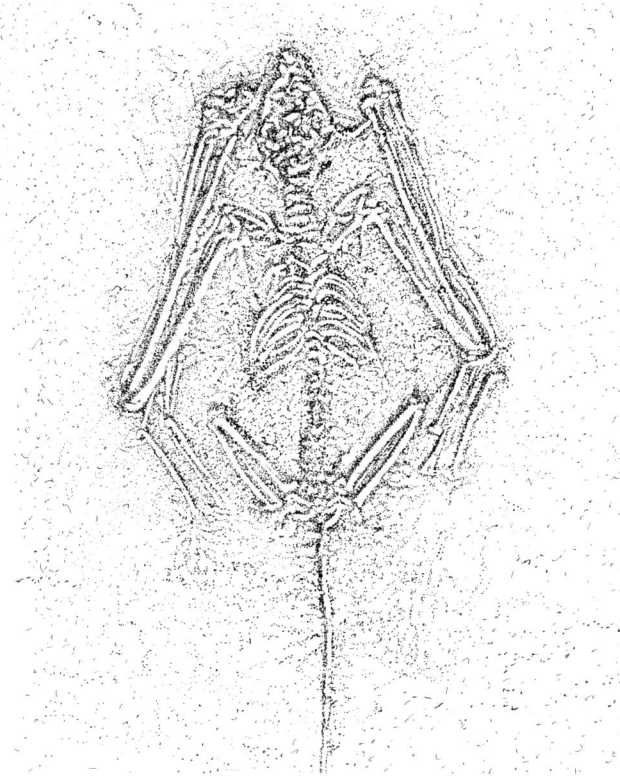

unanswered. When and how did bats evolve? What were the driving forces behind their evolution?

When did bats evolve?

If bats were advanced bats 50 million years ago, then they must have evolved earlier, but how much earlier? Certainly some time was needed to turn a small flightless mammal into at least four bat families, pushing the date back to 60 million years ago at the very least. Beyond that, the evidence becomes tantalising but tenuous. A fossil noctuid moth egg was found in 75 million year-old rocks in the United States. All known living and extinct noctuid moths have hearing organs, and these 'ears' almost certainly evolved specifically to listen for approaching predatory bats. The only logical conclusion is that echolocating bats fed on moths 75 million years ago. This and other evidence puts the origins of bats as far back as 100 million years ago. It is almost certainly the case that bats were among the small mammals that witnessed the demise of the dinosaurs 65 million years ago.

How did bats evolve?

Ask a zoologist and you will be told that bats started life as small, tree-living insectivores that jumped from branch to branch and tree to tree. They first developed gliding membranes and ultimately evolved powered flight (Fig.

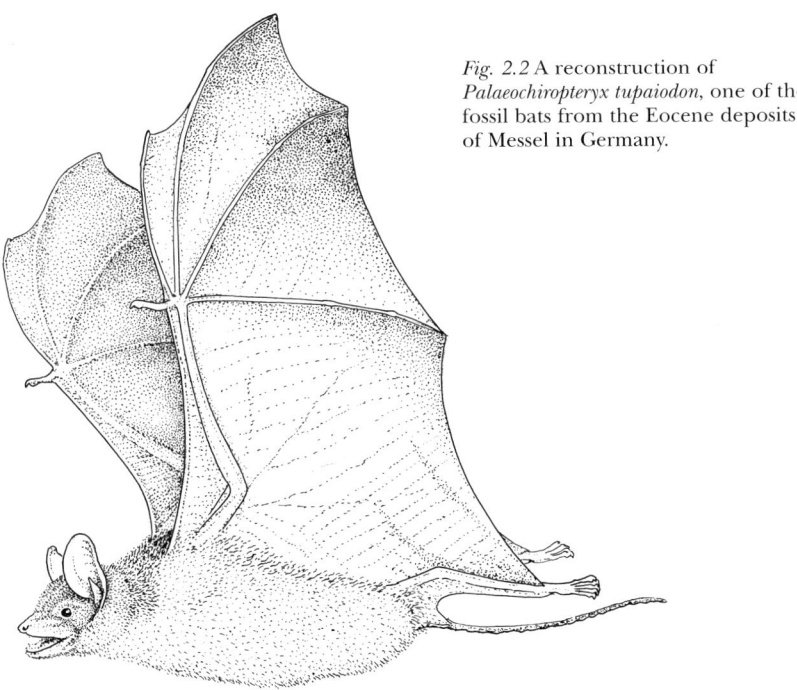

Fig. 2.2 A reconstruction of *Palaeochiropteryx tupaiodon*, one of the fossil bats from the Eocene deposits of Messel in Germany.

2.2). This answer is almost certainly correct, but there is no direct evidence to support it. There are no fossils of these intermediate forms, nor even of early bats. However, indirect evidence is all around us. First of all, becoming a poor glider is obviously relatively easy: gliding has evolved in many animals, including some very unlikely candidates such as snakes, frogs and fish. Next, the world is full of small insect-eating mammals that live in trees. Most of them get around by climbing and jumping, and many have gliding relatives. The evolution of skin flaps between fore and hind limbs produces a simple wing. Gliding has evolved many times among mammals, and some are very adept. In some mammalian gliders, notably the Dermoptera or colugos (misleadingly called 'flying lemurs'), the webbing extends between the fingers. A bat-like wing could evolve from this by simple elongation of the fingers. Finally, the early bat has to learn to flap its wings. Initially the flight muscles would have been unable to produce the sustained power needed for prolonged level flight. But gliding made this unnecessary, and true flight could have evolved at a leisurely pace.

What were the evolutionary driving forces?

We can only speculate on this question, but that does not make it any less interesting. Since flight requires major physiological and anatomical modifications, and demands considerable energy expenditure, it must offer an animal significant adaptive advantages. That flight does give a competitive edge in the fight for survival is evident from the very success of modern flying animals. Powered flight is found in insects, birds and bats. If success can be measured in terms of abundance, diversity and distribution, then all three can be said to be very successful. It is not difficult to produce a list of the advantages conferred by flight. Flight is expensive by the second, but, because it is fast, it is inexpensive by the kilometre. It therefore allows bats to cover long distances inexpensively and quickly. Bats can commute long distances between roost and foraging site, have large foraging areas, and make long seasonal migrations. Flight gives them new food resources, largely free from competition with other animals, and allows them to escape from predators.

Bats fly at night, probably because their ancestors were nocturnal, as are many modern small mammals. Bats are by no means blind (many have excellent night vision) but night flying is not an easy task, and echolocation evolved as an effective night-time orientation system. It probably evolved in parallel with flight, both systems being refined by natural selection at the same time. Although little studied, many other small mammals emit high frequency sounds, and some may echolocate after a fashion.

Bats are divided into two suborders, the mega- and microchiroptera. The megabats, as they are often known, are the Old World fruit bats or flying foxes. Confined to the Old World tropics, the 175 or so species feed exclusively on plant products – fruit, flowers, nectar and pollen, and sometimes leaves. All other bats are microbats, a widespread group that evolved from insectivorous ancestors, but which is now very diverse in its ecology and feeding habits. Figure 2.3 shows representatives of the two suborders.

It has long been assumed that all bats evolved from a common ancestor: in other words, bats evolved only once. It has recently been suggested that megabats evolved independently, some time after microbats, and from a primitive primate ancestor. Bat experts are divided on the issue, and as quickly as evidence is found to support one viewpoint, new evidence is found to oppose it, or support the

Fig. 2.3 Pteropus giganteus (left), a typical megabat, and *Nyctalus noctula* (below), a typical microbat.

other. It may be some time before there is a definitive answer, but common ancestry currently has the upper hand. The implications are considerable. If megabats did evolve independently, then the similarities between megabats and microbats are not the result of common ancestry, but of convergent evolution. In other words, the two groups must have solved the challenges of powered flight in remarkably similar ways, hence their striking similarities.

Diversity

Bats have had over 50 million years to diversify, and they have made an excellent job of it. The rest of this chapter is devoted to looking at this diversity. Since British bats make up just a tiny part of this diversity, to reveal its full splendour I have to tour the globe. This diversion is not only interesting, but essential to a true understanding of bats.

   Over the long history of life on earth, whole families, orders and even phyla have gone through cycles of boom and bust. Many have had their period of glory before giving way to the next group. Major cycles of extinction and adaptive radiation have their origins in widespread and even global changes in climate, brought on by shifting tectonic plates, ice ages, and even the impact of meteors or comets. Minor cycles may be controlled by more subtle ecological factors. All can be influenced by chaos (in the mathematical sense) and are often therefore beyond explanations based on intuitively straightforward cause and effect.

Feeding habits

Just what triggered the radiation of the bats is not known, but one mechanism seems most plausible. Around 100 million years ago, the flowering plants began their rise to prominence, and by 70 million years ago many modern plant families were already established. Insects diversified and multiplied alongside them, providing an abundant and varied food source for early insectivorous mammals, and this may have catalysed the radiation of the insectivorous microbats. Insect diversity is reflected in the bats that feed on them and, although most bats are generalists, many favour particular prey, taken in a particular way. Most feed on the small and abundant insects that can make the evening so uncomfortable for us even in Britain, such as the clouds of tiny biting midges that emerge at dusk. These bats feed on the wing, catching and eating an insect every few seconds when prey densities are high. Even here there is room for specialisation. Some feed out in the open, such as the streamlined, long-winged, fast and efficient fliers of the genus *Taphozous*, the tomb bats of the Old World tropics (Fig. 2.4). Even their snout and ears are shaped to minimise wind resistance. Some of the free-tailed bats (for example, *Tadarida* and *Chaerephon*) have stiff, forward-pointing ears that look as if they might be sufficiently good aerofoils to generate lift! The British bat that comes closest to this way of life is the large, streamlined noctule, *Nyctalus noctula*. Others skim just above the water taking emerging insects from the air, or use their large feet to trawl insects from the surface itself, like our Daubenton's bat, *Myotis daubentonii*. Others feed in vegetation, and their slow and manoeuvrable flight, essential to negotiate dense foliage, is aided by short, broad wings. Many species take insects from surfaces: the ground, leaves, tree trunks, etc., a strategy referred to as gleaning. Their prey may be large and sometimes incapable of flight. The California leaf nosed-bat, *Macrotus californicus*, will take large ground beetles

Fig. 2.4 Left: a tomb bat, *Taphozous mauritianus*, in flight. Right: the California leaf-nosed bat, *Macrotus californicus*, gleaning a beetle from a cactus.

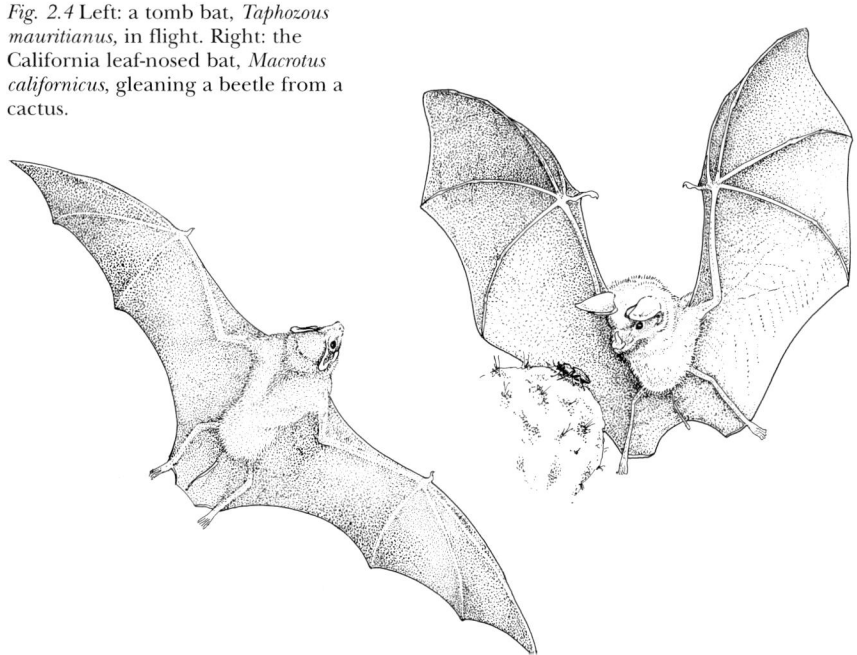

and crickets (Fig. 2.4), often hovering over its prey before pouncing.

Some of these gleaning species locate prey by listening for the sounds made by the insects themselves as they scuttle about. Many such species have evolved large ears for the purpose. Some rely entirely on their eyes on bright nights, and a look at *Macrotus californicus* reveals the large, forward-pointing eyes of a visual predator. The pallid bat, *Antrozous pallidus*, has long been one of my favourite bats, a status only enhanced by meeting it in person in the deserts of Arizona. This pale, white and yellow bat often feeds by gleaning, and, although little bigger than our own noctule bat, will take scorpions and giant, poisonous centipedes. Several British bats are gleaners, most notably the Brown long-eared bat, *Plecotus auritus,* and this too will often listen for prey, hunting without echolocation.

Some of the largest microbats (large means a mere 30–150 grams in the bat world) have evolved a taste for vertebrate prey: about 13 species are confirmed carnivores. The fish-eating bat, *Noctilio leporinus,* of South and Central America, gaffs small fish from the water in the same way that some of its smaller cousins take insects. It has long, sharp claws to grasp the fish, but it immediately transfers its prey to its mouth before finding a perch on which to eat it. The fish, or the ripples they cause on the water surface, are detected by echolocation. One population of *N. leporinus* has been studied in some detail. In the wet season, when insects are abundant, these comprise most of its diet, but in the dry season it switches to fish and crabs. Large piles of empty fiddler crab claws can be found under its night roosts. Its close relative, *N. albiventris,* and the vesper bat, *Myotis vivesi,* may also supplement their insect diet with fish, but there are no confirmed records of the British trawler, Daubenton's bat, trying the same trick.

The fringe-lipped bat, *Trachops cirrhosus*, of the New World, is another insectivore turned carnivore, and often consumes large numbers of small frogs. What is particularly fascinating about this bat is that it uses the mating calls of male frogs to pinpoint its prey. It can not only tell the difference between edible and inedible frogs, but concentrates on the bigger males that make bigger meals. Larger, fitter males are able to make more complex calls to attract females, and the bats have learnt this. *Trachops* is one of several carnivores belonging to the wonderfully diverse Phyllostomidae, the spear-nosed bats of the neotropics, which between them take amphibians, reptiles, birds and mammals in addition to insects and plant products. The large (150 grams) false vampire bat, *Vampyrum spectrum*, may sometimes specialise on birds, and can be so successful that it only needs to feed during one night in three. The Old World also has its carnivores, the pale, blue Indian false vampire, *Megaderma lyra*, and the large and striking Australian ghost bat, *Macroderma gigas*, which has a wingspan approaching 1 metre. *Macroderma* has many of the characteristics common to carnivorous bats: short, broad wings that allow it to take off easily with large prey, and large, forward-pointing eyes and large ears, which may add passive sound and vision to echolocation as prey detection systems (Fig. 2.5).

Next is another group of phyllostomids, the vampires. There are three species, the common vampire, *Desmodus rotundus*, and the much rarer hairy legged, *Diphylla ecaudata*, and white-winged, *Diaemus youngii*, vampires. *Desmodus* now shows a strong preference for feeding on domestic animals, the latter two feed primarily on wild birds. *Desmodus* colonies of up to 100 bats roost in caves and hollow trees. Emerging at dusk to feed, they locate their prey using all their senses, including smell. They are remarkably agile and often approach their chosen prey by crawling along on the ground. Having located areas of warm skin rich in surface blood vessels, using heat-sensitive cells on the nose, they make small, shallow incisions with their sharp incisors and lap up the flowing blood. Anticoagulant in their saliva keeps the blood flowing. This is now synthesised for medical use and appears to have some advantages over other clinical anticoagulants. A 35-gram bat can drink 25 ml of blood in one meal, hence the specific name: *rotundus*! Vampire bats can usually go only three or four days without food, but *Desmodus* has evolved a behaviour to help

Fig. 2.5 Macroderma gigas, the carnivorous Australian ghost bat.

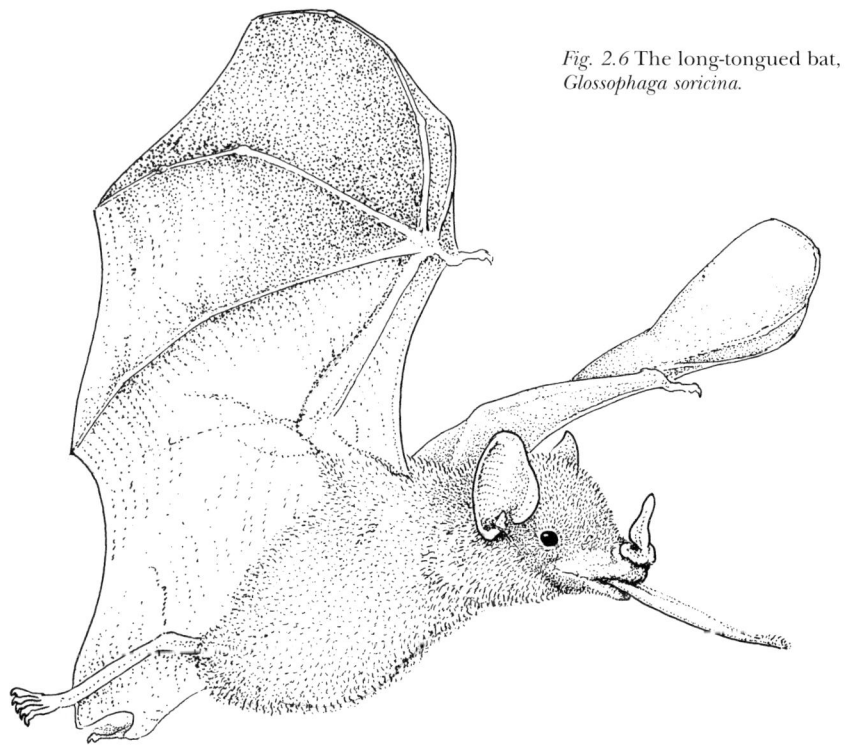

Fig. 2.6 The long-tongued bat, *Glossophaga soricina.*

counteract the sometimes ephemeral nature of its food supply. In the roost, bats form lasting social bonds, with the strongest bonds being established between pairs of often unrelated individuals. If a bat fails to get a meal on a particular night, it can beg food (regurgitated blood cells) from its partner. On a later occasion, the favour will be returned. I find it very fitting that in a bat which has such a bad public image, biologists have found a rare example of altruism in nature, a behaviour typically attributed in the public mind to more socially acceptable animals.

Finally amongst the microbats there are the frugivores: phyllostomids with an insectivorous ancestry. There is little the phyllostomids do not eat, and their evolution to fill a wide range of neotropical ecological niches is a rich and interesting lesson in adaptive radiation, which is the evolution of a variety of species from a common ancestor to fill new, vacant ecological niches. These bats presumably developed a taste for fruit, nectar and other plant products by accidentally eating them when taking insects from fruit and flowers. In the modern phyllostomids all possible stages of the transition may be seen, from the omnivorous spear-nosed bat, *Phyllostomus hastatus,* through to the specialised hovering nectar feeders such as the long-tongued bat, *Glossophaga soricina* (Fig. 2.6). In between are frugivores that occasionally take insects, such as the tent-making bat, *Uroderma bilobatum.*

Why were the phyllostomids able to diversify in such a way? Probably an accident of birth – they evolved in the vast neotropical forests, a region of very high

plant diversity with few competitors. Since a given patch of forest typically produces a wide variety and reliable supply of fruits, specialisation has been made
easy. The absence of the megabats, which have always been confined to the
Old World, may have been important too.

Next we come to the megabats, the microbats' vegetarian cousins. The
majority of megabats are generalist frugivores. Palaeotropical forests are less
diverse than their neotropical counterparts, and the fruit supply is patchy, in
both time and space. Megabats have to be prepared to travel some distance to
feed, and must be less choosy about what they eat than neotropical frugivores.
However, there are some specialists among them, including some small nectar-
feeders such as the long-tongued fruit bats, *Macroglossus* species.

Between them, over 250 species of Old and New World bats feed on plants
from over 300 genera, and over 500 plant species are pollinated by bats. We are
only now beginning to appreciate fully how important bats are as pollinators
and seed dispersers. They play a key role in the maintenance of tropical ecosystems, and are critical to the natural spread of tree species, and to forest regeneration after natural or human clearance.

Perhaps one of the most unexpected relationships is that between a small
nectar-feeding bat and the spectacular plants of the Sonoran Desert in the
south-western United States and Mexico. After many years of slow growth, an
Agave palmeri puts forth a huge flower spike, over 5 metres high, which produces vast quantities of nectar 'specially formulated' for bats. The major bene-
factor is the long-nosed bat, *Leptonycteris curasoae*, which in turn repays the
agave, as its prime pollinator. The crushed succulent leaves of this agave are
used to produce tequila.

The giant saguaro cactus, star of many western movies, is also heavily dependent on *Leptonycteris* (Fig. 2.7). The saguaro has been described as a 'keystone'
species of the region: a plant around which much of the ecology of the desert
revolves. The survival of many ecosystems may depend upon the fortunes of a
small number of keystone species, and it is easy to see that many animals rely
on the saguaro for food and shelter. In the past, disturbance due to the
removal of saguaro for gardens and city developments, and the wholesale
removal of agaves for illegal tequila production, has upset the fragile natural
balance of the desert. Without careful management the consequences could
be serious: too few food plants, and the desert cannot support a viable bat population. Without the bats, the few plants left cannot survive, and with their loss
the desert dies. Fortunately, the beauty, subtlety and fragile diversity of the
Sonoran Desert is now appreciated and better protected.

On the other side of the world is another fascinating relationship between a
bat and a plant. The short-tailed bat, *Mystacina tuberculata*, endemic to New
Zealand, still flies, unlike many of New Zealand's unusual birds, but it also
spends a great deal of time on the ground. To suit this way of life it has thick,
leathery wings, the delicate tips of which can be tucked into pouches when the
bat is on the ground, and it has extra talons on its short, thick toes. It has been
known to use these talons to burrow into rotten trees to roost. This odd little
bat has a mutually beneficial relationship with New Zealand's equally unusual
and endangered parasitic plant, the wood rose, *Dactylanthus taylorii*. The large
flower clusters of the wood rose open directly on the soil surface around the
trees on which it feeds, and these flowers produce a great deal of nectar. This
had long been a mystery, since it implied a large, vertebrate pollinator, but few

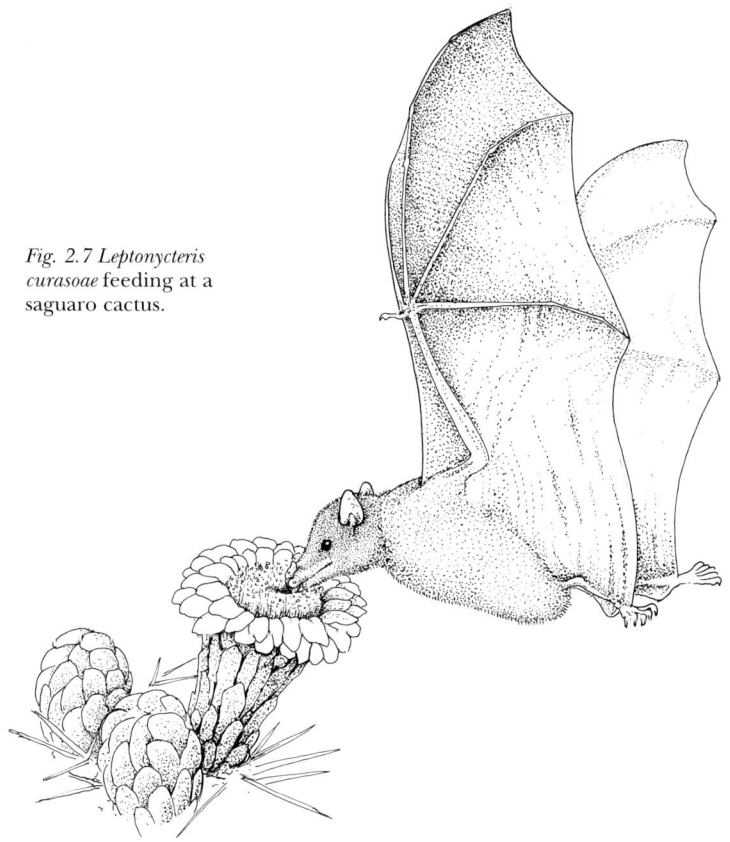

*Fig. 2.7 Leptonycteris
curasoae* feeding at a
saguaro cactus.

candidates suggested themselves. Another mystery was just what was eating the
flowers of *Dactylanthus* before it could be pollinated? This was important, since
it might be the key to the plant's serious decline. The forest service set up all
night time-lapse video and caught the flower thieves: introduced rats and pos-
sums. To their delight they also caught the pollinator: short-tailed bats were
making up to 40 visits each night to drink the nectar. Both species appeared to
have a tenuous hold on life, but the discovery of this unique relationship gives
new hope to both. New colonies of short-tailed bats have been found close to
known populations of wood rose and management plans aim to safeguard
both.

In 1883 Krakatau, a small island between Java and Sumatra, was virtually
destroyed by a volcanic eruption of enormous proportions. All life on the rem-
nant islands was wiped out, marking the start of a unique natural experiment
in recolonisation. In the century since the eruption, a complex ecosystem has
established itself on the islands, and bats have played an important role in this
recolonisation. They have carried seeds of many plants, including keystone
trees, from the mainland and dispersed them throughout the Krakatau archi-
pelago. Bats are often the major, or even the only, vertebrate pollinators and
seed dispersers on Pacific islands. Unfortunately they are also amongst the

most threatened animals and their extinction could have serious conse-quences on island ecology.

The relationship between bats and plants is a long one and the evidence for their coevolution is easily seen. Many plants pollinated by bats produce flowers that open only at night. In many cases the flowers open for just one night and fall before dawn. Flowers are typically white and often emit pungent odours, making them conspicuous to megabats with their good night vision and acute sense of smell. One neotropical plant attracts echolocating, fruit-eating phyl-lostomids with shallow flowers specifically shaped to return powerful echoes. Flowers with deep bowls fit snugly round the muzzles of some bat species and have their anthers strategically placed to transfer pollen to the bat. The nectar produced for bats can be very different in composition to that produced for insects: *Agave palmeri* produces nectar with a significant lipid content. Adaptations are evident in the bats too: elongated muzzles, loss or reduction of teeth, and long tongues to collect the nectar. The bats get much of their pro-tein from eating pollen and some species have especially scaly hair around the head and neck to hold pollen. Behaviour has also adapted to patterns of nec-tar production. *Leptonycteris curasoae* feeds in flocks, visiting agaves in such a way as to get the most nectar for the least effort – optimal foraging. The flock moves from one plant to another, staying only if each bat regularly gets a good 'mouthful' of nectar as it dips into a flower. When a plant's supply of nectar begins to run low they move on to another plant, but never to one already vis-ited that night. The nectaries are half-full again the next night, and almost full the night after, so the bats can begin to exploit these plants again very soon.

Roosting habits

Bats spend more time roosting than on any other activity. The roost provides protection from the elements and from predators. It supplies improved oppor-tunities for mating, rearing young, exchanging information about food sources, and cheaper thermoregulation. Different bats derive different benefits in dif-ferent seasons from their roost, and the enormous diversity in ecology, habitat and climate has led to an equal diversity in roost site and roosting ecology. There are species that habitually roost singly and species that roost in colonies of several millions. There are bats that return year after year, generation after generation, to the same durable roosts in caves and crags, others that move almost nightly. I will illustrate some of this variety with a few examples and show how roost site and roosting ecology are related to the life history of the species.

The colonial cave dweller is what many people consider to be the archetyp-al bat. There is no shortage of examples, but it should be said that cave-dwelling bats are probably in the minority. This is not surprising given the rel-ative scarcity of caves in many parts of the world and the unsuitability of many of them for bats. However, where suitable caves do occur, bats can be at their most spectacular.

The tiny (6 gram) Mexican free-tailed bat, *Tadarida brasiliensis*, forms the biggest known colonies. Bracken Cave in Texas is home to some 5–20 million individuals, and each summer night these fast-flying bats emerge to feed on insects over a circle many kilometres in radius around the cave. Bracken Cave is a summer nursery roost. It was once thought that no mother could possibly find its own baby on returning to the cave, so it was assumed that suckling was indiscriminate. Careful studies have since shown that mothers do locate and suckle their own young. It is only the occasional milk thief that snatches a feed from another bat's mother. Many other species also form colonies of immense size: the leaf-nosed bat, *Hipposideros caffer*, in Africa and long- or bent-winged bat, *Miniopterus schreibersii*, in many parts of the Old World can total hundreds of thousands to over half a million. Several species may roost in the same cave, and a number of sites are home to mixed roosts of over a million bats. Although different species will share the same cave, they usually roost separately, but closely related species may take up adjacent sites. At the other end of the scale, many species roost in small groups or even singly. Some of the large, carnivorous members of the Megadermatidae and Phyllostomidae roost in small family groups in caves in the company, surprisingly, of other bats that are potential prey. Several megabat species also make use of caves. Most of these rely on good night vision to find roost sites relatively close to the entrance. Only *Rousettus* species roost deep in caves and they have evolved a unique form of echolocation to enable them to navigate in total darkness. In contrast to microbats, which generate their echolocation calls in the larynx, *Rousettus* click their tongues to produce a very different call that nevertheless appears to be effective. Interestingly, this form of echolocation is similar to that used by the Old World cave swiftlets and the New World guacharo, unusual cave-living birds.

A single cave can provide a wide range of substrates and microclimates through the seasons, leading to a complex and constantly changing bat community. From large caves through small caves and rock shelters to crevices between boulders and cracks in the rocks themselves, all are home to bats. Crevice-dwelling bats are well known in desert regions: the pallid bat, *Antrozous pallidus*, makes widespread use of caves in the south-western United States. By moving around between cracks of different orientations and depths, they can avoid the worst extremes of desert temperatures. Some bats are visibly adapted to crevice dwelling. The 10–20-gram South African flat-headed bat, *Sauromys petrophilus* (Fig. 2.8), as its name suggests, has a flat head that enables it to crawl into the narrowest of rock crevices to roost. On the many rock inselbergs in the Namib Desert, exfoliating flakes of granite provide perfect roosts. Crevice-roosting relatives of *Sauromys* with similarly flat heads are found in other parts of Africa and in South America.

Roosts in plants are more widespread and diverse than rock roosts. Cracks, crevices and the hollow trunks of trees are probably home to more bats in more parts of the world, including Britain, than any other roost type. Although

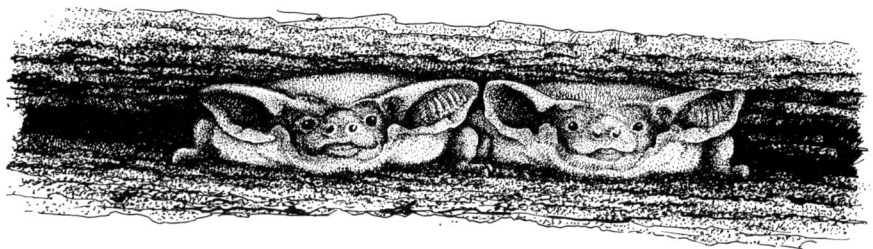

Fig. 2.8 The flat-headed bat, *Sauromys petrophilus*, in a rock crevice.

they are short-lived relative to most rock roosts, bats can use the same tree hole for many generations. More frequently, a colony of bats will have several tree-hole roosts and will move around them as the climate and local food supply demand. Roosts can harbour ectoparasites (small invertebrates that feed on blood, exfoliating skin or other parts of bats) so it pays bats to abandon roosts at intervals to let the parasites die off. Peeling bark provides a ready supply of temporary roosts. Hollow trunks are a common feature of many tropical trees and a single trunk can house large colonies of several species. It has even been suggested that hollow trunks have evolved to attract bats and other animals, since their guano provides the tree with valuable nutrients in poor tropical soils. In the absence of trees, other large plants are used. In the Sonoran Desert of the south-western United States and Mexico, holes in tall columnar cacti such as the saguaro, often made by woodpeckers, are used by several species. In warmer parts of the world, bats will roost on the outside of trees. The most visible are the large and often noisy camps of Old World megabats of the large genus *Pteropus*, that roost by the thousand in the open tops of larger trees, defoliating the branches. Being largely safe from predation, they have no need to hide. Smaller megabats are often much less conspicuous, roosting in smaller groups, hidden amongst the leaves or against the trunk. Many microbats also roost in the open. The tiny South American proboscis bat, *Rhynchonycteris naso*, clings to the bark of trees that overhang streams, where its cryptic fur passes it off as a piece of peeling lichen. The beautiful butterfly bat, *Glauconycteris superba* (Fig. 2.9), has strikingly patterned fur and reticulated wings that may be an adaptation to camouflage the bat. The wings can resemble dead and wrinkled leaves and the whole bat a dying flower as it hangs amongst the foliage, but this is still mere speculation.

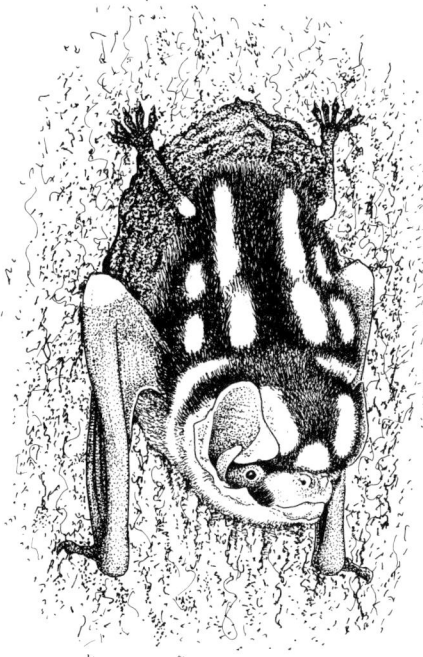

Fig. 2.9 The butterfly bat, *Glauconycteris superba*.

Bats use some very unusual roosts. The two species of bamboo bat, *Tylonycteris pachypus* and *T. robustula* of India and Southeast Asia, roost inside the stem or culm of giant bamboo, but they need assistance getting in. A large chrysomelid beetle pupates in the culm and has to eat its way out. This large beetle leaves an elongated hole big enough for the bats to enter. They grip the smooth sides of the culm with the aid of fleshy 'suction pads' on their wrists and ankles. The New World disk-winged bats, *Thyroptera* spp., and the sucker-footed bat of Madagascar, *Myzopoda aurita*, have very well-developed pads that they use to grip the smooth, inside surfaces of furled leaves. These can only be temporary roosts, since as the leaves unfurl the bats must move on.

Suction pads are used by no less than 11 vesper bats. One of them, *Myotis bocagei*, has been known to roost inside water arum flowers. Rolled banana plant leaves have become very popular with several species and plantations may have altered the abundance and distribution of some of these bats. Suction pads are not a prerequisite for roosting inside leaves: the banana bat, *Pipistrellus nanus*, has no need of them.

At least 17 species, both microbats and megabats, have to some degree or other learnt to make their own roosts by strategically chewing and manipulating leaves. This behaviour has evolved independently at least three times in different parts of the world. It is most widespread amongst the neotropical phyllostomids and the usual practice is to chew through the veins and ribs that support large and often multi-lobed leaves. This causes the leaf to collapse around the bats like a tent, protecting them from the weather and the eyes of predators. The tiny white bat, *Ectophylla alba* (Fig. 2.10), takes on a cryptic green colour when lit through the roof of its tent, and it is tempting to suggest that it may have evolved a white coat for camouflage. Depending upon the plant and the construction method, the tents may last from days to weeks, and at any one time a small group of bats may be using several tents.

Several species are known to use the homes of other animals as their roost. The cave bat, *Myotis velifer*, has been found roosting in old cliff swallow nests in the south-western United States. In West Africa, woolly bats, *Kerivoula* spp., have been found in the large webs of a colonial spider. Round-eared bats, *Tonatia silvicola*, excavate roosts in suspended termite mounds in South America, and slit-faced bats, *Nycteris* spp., roost in old aardvark burrows in Africa.

Fig. 2.10 The Honduran white bat, *Ectophylla alba.*

Humans have done much to destroy or disturb the natural homes of bats and this has undoubtedly contributed to the serious decline of many species. However, the relationship has not been entirely one-sided. Bats can be very adaptable and throughout the world they have adopted man-made structures as roost sites. Mines and tunnels are used by cave-dwellers, cracks in buildings resemble rock crevices, timber roofs offer roosts not unlike the varied cavities found in trees and grass huts and roofs can be substitute foliage roosts. Bridges, from ancient stone structures to the latest concrete designs, are a favourite roost of many species, from Daubenton's bat, *Myotis daubentonii*, in Europe to the Mexican free-tailed bat, *Tadarida brasiliensis*, in the New World. Perhaps the most publicised roost in the world is the Congress Avenue Bridge in the centre of Austin, Texas. Over one million 'free-tails' roost under the bridge, and Bat Conservation International, which is based in Austin, have used it to great advantage in their conservation campaigns. Man-made roosts may have allowed some bats to extend their ranges. In Europe, the Northern bat, *Eptesicus nilssoni*, takes advantage of buildings beyond the Arctic Circle during the short but productive summer. Was this bat able to thrive this far north before humans moved in?

At the beginning of this section one of the advantages of communal roosting that was mentioned was the exchange of information. Almost every night in summer, all the bats in a roost leave in search of food. In many instances each bat may find an elusive night's feed through its own efforts, but there is increasing evidence that bats learn from each other and even co-operate in finding food. Many bats, particularly lactating females, return to the roost one or more times through the night to suckle their young. A bat that has failed to find a good insect meal can follow a roost mate from the roost to its foraging site in the hope of getting a good meal. How is this known for sure? The entrance to a roost of evening bats, *Nycticeus humeralis*, was cleverly rigged so that ringed, colour-coded bats automatically weighed themselves and took their own photograph as they came and went. Hungry bats left with well-fed bats and came back full. Some bats leave the roost together and go to the same foraging site, calling to maintain contact. A group of bats is more likely to find elusive food than a single bat and the successful individual can emit calls to attract its roost mates. This sort of behaviour has been little studied in British bats, so there is much left to investigate.

Living together has its rewards, but it also brings drawbacks. One reason for bats to move roost regularly may be to escape from infesting ectoparasites. Another is to escape from predators. An owl or hawk (or a domestic cat) can quickly learn to take advantage of food on the wing that appears reliably at dusk every night. There is evidence from studies of free-tailed bats in southern Africa that bats will change roost and vary their emergence time to reduce predation risk. In Britain, the emergence of pipistrelle bats in non-random bursts is most likely a mechanism to confound predators.

Reproduction and life history

The life history strategies of bats are nothing if not diverse, but in one respect bats are all very similar. Although small mammals, their life cycles are far more typical of much larger ones. Like virtually all subjects in biology, mammalian life history strategies are a complex continuum. However, as mammals get bigger there is a clear trend towards longer life expectancy and lower fecundity.

Small mammals tend to get little care after they are born, gain independence and sexual maturity at an early age, breed young and often and subsequently die young. Small rodents and insectivores, for example, reach maturity in a few months and produce litters of upwards of four young every three or four weeks, but are lucky to reach their first birthday and very unlikely to see their second. Bats, being small, might also be expected to live fast, furious and short lives and leave behind many offspring, most of which will die before they can breed. But they do not. Bats live much longer than would be expected for their size, and the vast majority of species produce a single baby (Fig. 2.11), once or perhaps twice a year. Twins are uncommon, triplets or more exceedingly rare.

Gestation is comparatively long and young are born large, so the time to weaning is short. Mortality is necessarily low, given the low birth rate. Sexual maturity can be reached in a few months, but is typically one to two years. Even the smallest bats can live for five or more years and there are many reliable records of wild bats living for more than ten years, and a few for more than 30. Just why bats live so long is a complex question. It was once thought to be related to their ability to hibernate. The slowing down of physiological processes during hibernation was presumed to slow ageing in some undefined way. However, bats from only two families, the Vespertilionidae and the Rhinolophidae, are true hibernators, yet most of the bats from the other 15 families of microbats and the megabats are all long-lived. Low fecundity may be a consequence of flight in mammals and this in turn may necessitate longevity. This will be discussed in due course (p. 61), but it does not answer the question of how they live so long, perhaps just why.

The details of life history and reproductive cycles will be discussed further in Chapter 3. Here I will describe some of that diversity mentioned at the beginning of this section. Many bats are monoestrous, that is, they reproduce just once a year, when food is abundant, so that the mother can feed the growing foetus and then the baby before it is weaned. In temperate regions this means the spring and summer. In the tropics birth coincides with the food abundance of the rainy season. Tropical bats have the option of breeding more than once a year, if there is more than one rainy season or the season is a long one. Schreiber's bent-winged bat, *Miniopterus schreibersii*, a bat with a very wide geographical distribution, is monoestrous in temperate regions and polyoestrous in the tropics. Most tropical bats regularly breed twice a year, sometimes three times, although not all attempts are equally successful.

Although pregnancy and birth may coincide with periods of abundant food, mating and copulation may not. Bats have a num-

Fig. 2.11 Egyptian fruit bat, *Rousettus aegyptiacus*, with young.

ber of physiological strategies that enable them to mate and then delay either fertilisation or foetal growth until the conditions are right. This means that the timing of the sometimes complex and protracted mating process is not critical: bats can fit it in when 'convenient' in their busy lives.

Mating itself is a complex process driven by a number of ecological, behavioural and physiological factors and given the diversity of these factors in bats it is no surprise that mating systems are equally diverse. Evolution favours those males and females that have strategies to maximise the number of offspring they produce, and whose offspring survive long enough to reproduce themselves. But the best strategy for a male to adopt often depends upon the behaviour of the females. In turn, this may be determined to a large extent by what they feed on and how they forage. It is now generally believed that female social structure and ecology are often the major driving forces in the evolution of mating systems. Male mating strategy appears to evolve to make the best of the females' behaviour. These strategies usually fall into one of several basic categories. Monogamy (the partnership of a single male and single female for at least one breeding season) is rare in nature and occurs most frequently when the successful rearing of young requires the active participation of both parents. It has not been confirmed in many bats, but may be more common than in most other mammals. Another category involves the defence, by one or more males, of groups of females. If the females occupy small home ranges then the males can defend a territory that encompasses this range.

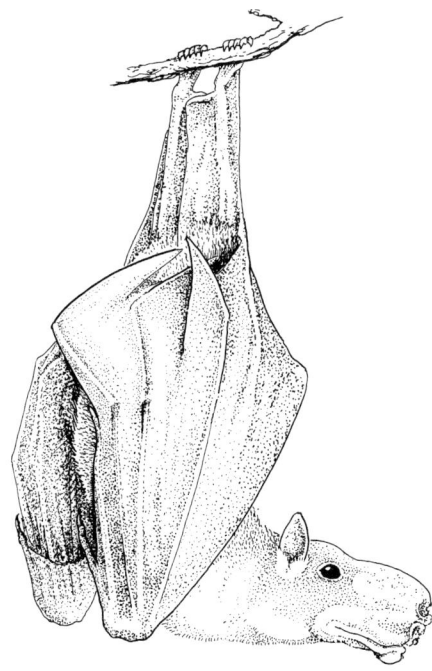

Fig. 2.12 Male hammer-headed bat, *Hypsignathus monstrosus*, calling to attract mates.

Alternatively, the females themselves may be defended if they form a group of stable composition even if their home range is large, since the males can simply follow the females around. Males can also defend areas of choice habitat, or simply good roosts, and mate with females that are attracted to these sites. Males can set up leks: small temporary mating territories from which they display and compete for passing females. The best known lekking bat is the hammer-headed bat, *Hypsignathus monstrosus* (Fig. 2.12). Males hang upside down in trees and produce loud and persistent honking sounds with the aid of a greatly enlarged larynx, as they flap their wings. Passing females select suitable males for mating, presumably on the quality of their honking. This is one of the few marked examples of sexual dimorphism in bats, where males and females of a species are noticeably different. Another good example is the crested free-tailed bat, *Chaerephon chapini*. The males of this species have a beautiful erectile crest of long pale hairs on the crown of their head. Finally, mating may be entirely promiscuous, taking place in a wide range of situations. Leks are not particularly common amongst bats, but all the other strategies are widespread and will be discussed later, in the context of the British species (pp. 64–69).

Hibernation

It comes as a surprise to many people to hear that true hibernation occurs in only two of the 17 microbat families. Hibernation is often seen as a defining feature of bats, probably for several reasons. First of all, both families in which hibernation occurs, the Vespertilionidae and the Rhinolophidae, are widespread and large. In fact the Vespertilionidae is easily the largest and the most widespread. Its 330 species are found everywhere but the Antarctic, much of the Arctic, and a few remote islands. Secondly, temperate bats, which make full use of hibernation, are the best studied. Comparatively little is known about the biology of the vast majority of tropical bats. It is almost certainly the case that bats evolved in the tropics and that hibernation was a late evolutionary step. Torpor, the ability to lower body temperature in a controlled way to save metabolic energy, would be of value even to tropical bats when food is scarce. Many tropical bats, particularly small insectivorous species, are able to enter torpor for short periods, although body temperature does not usually fall more than a few degrees. Size and diet are important in determining whether torpor is likely to be useful to a bat. Small mammals have larger surface area to body mass ratios: that is they have small bodies to generate heat and relatively large surfaces over which they lose this heat to their surroundings. Small mammals have a hard time staying warm. To generate heat, they need to eat, so if their food supply is low in energy, unreliable, or both, staying warm is even harder. It is these bats that benefit most from an ability to go into torpor. It is assumed that as bats invaded the subtropics and then temperate regions, torpor became progressively deeper and longer. Hibernation is simply long and profound periods of torpor, but torpor is a versatile mechanism used by temperate bats at all times of the year as part of their energy budgeting programme (pp. 48–53).

When bats hibernate it is important to choose an appropriate hibernation site or hibernaculum. Each species has its own preferred body temperature during hibernation, which matches its unique physiology and minimises the rate at which the bat uses its stored energy. This is usually somewhere between 1 and 12°C. It makes sense therefore to choose a hibernation site where the temperature is constant and close to this ideal, since less energy is then used by the bat in actively regulating its temperature. Good hibernation sites usually have a high humidity, since this reduces water loss by bats: this can be as crucial as energy loss. A hibernation site also needs to be safe from predators, since a torpid bat is particularly vulnerable. Even seemingly benign disturbance, for example, in the form of visiting scientists, can cause arousal and the loss of vital energy stores.

For many species, hibernation conditions are so precise that hibernacula are rare and bats may travel very long distances to sites used for many generations. A single cave in Vermont once held 300,000 hibernating little brown bats, *Myotis lucifugus*, estimated to be the entire population of the surrounding

22,000 square kilometres. Ninety per cent of the entire south-eastern United States population of grey bats, *Myotis grisescens*, almost two million bats, hibernate in just three caves, 75 per cent of them in just one of these. In Britain, evidence suggests that species such as Natterer's bat, *Myotis nattereri* (Fig. 2.13), gather at suitable sites from a considerable catchment area.

Migration

Some bats hibernate when food is short, but there are other options. A surprising number of bats are migratory. Many tropical and subtropical bats migrate to fol-

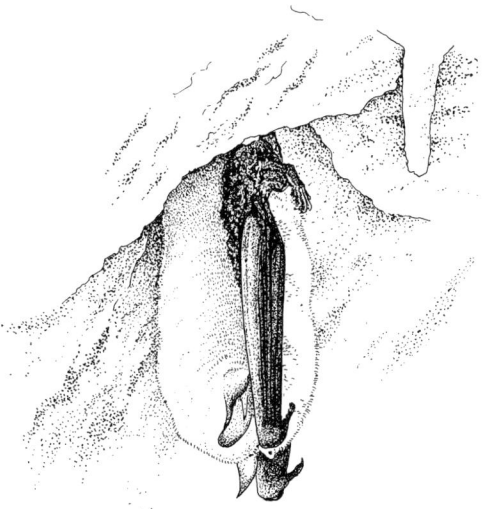

Fig. 2.13 Hibernating Natterer's bat, *Myotis nattereri*, hanging from one foot.

low sources of food. The nectar-feeding phyllostomid *Leptonycteris curasoae* migrates north and south along the Pacific coast of Mexico and into the deserts of California and Arizona, following the flowering of desert cacti and other succulent plants. Some temperate species migrate to the subtropics to find food. In North America, free-tailed bats and several vesper bats such as the hoary bat, *Lasiurus cinereus*, migrate south to Mexico and remain largely active throughout the winter. Some populations of the same species migrate shorter distances to hibernation sites. Some of the species undertaking short migrations to hibernate may even fly north rather than south, the journey being dictated by location of suitable hibernation sites rather than general climatic trends. Some populations of the grey bat, *Myotis grisescens*, fly from summer roosts in northern Florida to hibernation sites in Tennessee. In Europe, migration is invariably to hibernation sites and is typically south-west in autumn and north-east in spring, although short 'migration' flights can go in all directions. Migration distances can exceed 2,000 kilometres. Both large and small bats undertake these migrations, such as the noctule, *Nyctalus noctula*, and Nathusius' pipistrelle, *Pipistrellus nathusii*. Migration obviously requires considerable navigational skills, but very little is known about the mechanisms involved. Bats can orientate themselves by sunset glow, and vision may be important in other ways, but this is still an unexplored area.

3

The Biology of Temperate Bats

British bats are typical of temperate species throughout the world, and much of what is known about them has come from studies of the same or similar species in Europe and North America. This chapter describes in some detail aspects of their biology that are important to an understanding of their natural history. Wherever possible British species will be used to illustrate not only how bats function, but how evolution through natural selection has 'designed' them to work in that particular way. The 'designs' of evolution are referred to for convenience, but this is shorthand for the products of evolution by natural selection. The designer has no notion of the future. An animal may look well designed, in that it performs its various tasks well, but this design is the product of a slow and wasteful process of trial and error. Natural selection demands that bats are continually evolving towards some optimal solution to the problem of survival and reproduction in a competitive, hostile and changing world. The optimum is never achieved for several reasons. The environment is constantly changing, evolution works slowly, and the animal is often trying to catch up with its environment. Each animal also carries with it the baggage of its past history: the starting material limits the designs possible and new designs have to be built seamlessly on old ones. Finally, most designs are compromises: animals have to perform many tasks and few can be masters of all trades. However, the solutions are often surprisingly close to the best that we can construct ourselves. In addition to revealing the inner workings of bats, this chapter shows some of the successes and the limitations of natural selection. Each aspect of bat biology is dealt with separately and the essential details covered. Chapter 4 builds on this foundation and draws all these components together.

Flight

The single most striking feature about bats is that they fly (Fig. 3.1). They are the only mammals capable of powered flight. This has had an all-pervasive influence on their lives: their anatomy, physiology, behaviour and their ecology influence and are influenced by their ability to fly. Being able to fly is a *good thing*. Look at the animals that fly: insects, birds and bats are all abundant, diverse, widespread and clearly successful groups. What do they get out of being able to fly? To put it simply, flight is fast and inexpensive. The energy expended each second is high, twice as great as running, but because flight is fast, the cost per kilometre flown is low, as little as one fifth of the cost of moving over the ground. Even if the cost is ignored, just being fast has its advantages. Flight allows animals to live in one place and feed in distant places. It allows them to search over wide areas for food and take advantages of food sources unavailable to other animals. It can give an animal a unique viewpoint from which to search for food. It allows animals to migrate quickly over long distances as the seasons change. Migrating animals can cross deserts, mountains and oceans that are difficult or impossible barriers to terrestrial animals.

Flight allows animals to escape from predators and to rest, sleep, feed or breed in relative security. When you see all the advantages, you find yourself asking the obvious question. If flight is such a good thing, why don't all animals fly? The answer to that is that flying is not easy, or at least evolving the ability to fly is not easy. It requires a very high specification plan: design tolerances are low, and compromises must be made in other functions to guarantee success.

From climbing to gliding to flying

Just how vertebrates evolved flight is a very controversial and complex issue. Did birds evolve from fast terrestrial bipeds that jumped into the air, or from ancestors that climbed trees and cliffs, to jump off and glide? The anatomy of a bat suggests that its ancestors were very unlikely to have been sprinters. However, there are many small, tree-climbing mammals that have gliders amongst their kin, such as the Dermoptera ('flying lemurs'), flying squirrels and the marsupial sugar gliders. In fact, every other class of vertebrate has its gliders: fish, frogs, lizards and even snakes.

The ancestral bat was probably a small, nocturnal, insectivorous mammal that lived in trees and made its living by running and jumping from branch to branch. As in many modern gliders, the development of a flap of skin between the fore and hind limbs slowed down the glide, increased stability and manoeuvrability, and decreased the glide angle to allow the animal to glide further. A critical step would have been the evolution of a gliding membrane between the fingers and the extension of the fingers to create a bat-like wing. A wing designed for gliding is also capable of being used for flapping flight and even very weak and shallow flapping will assist the glide. Powerful, flapping flight can then evolve by a progression of small changes in the anatomy and physiology of the glider.

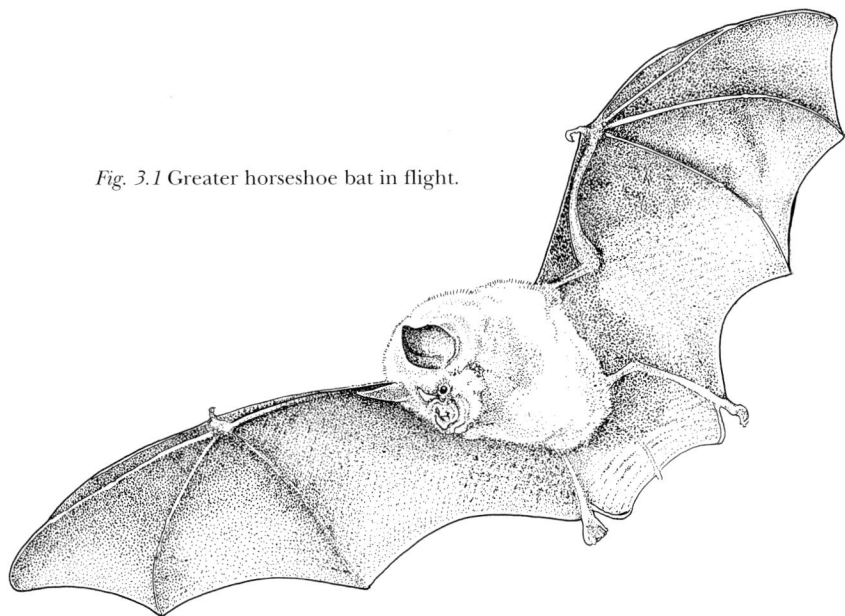

Fig. 3.1 Greater horseshoe bat in flight.

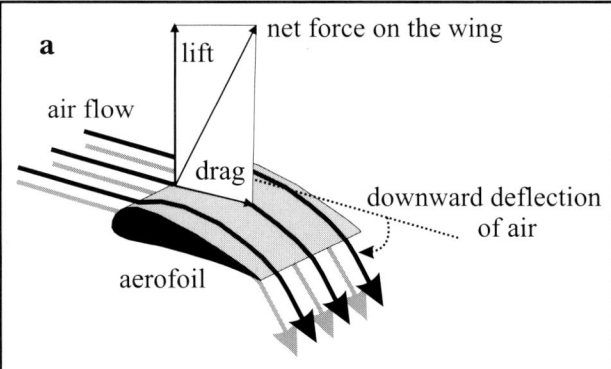

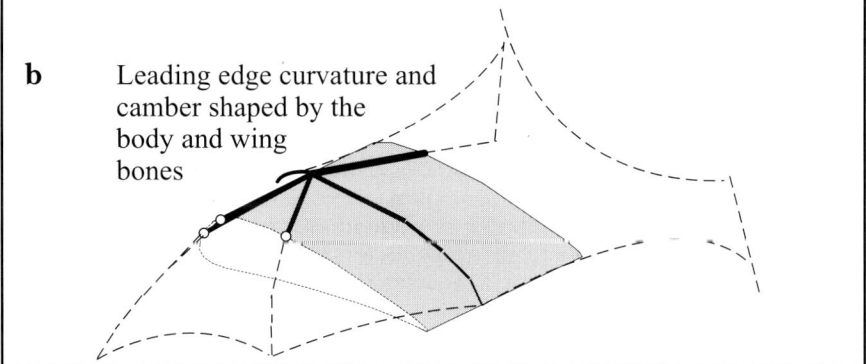

Fig. 3.2 (a) An idealised aerofoil and the aerodynamic forces around it. (b) How a bat's wing works as an aerofoil.

Some simple aerodynamics

Gliding and powered flight are made possible by the simple aerofoil geometry of the wing. In cross section an aerofoil has a blunt or rounded leading edge that faces into the wind. From this edge the aerofoil tapers to a pointed trailing edge and curves gently downwards. Figure 3.2 shows how a bat's wing conforms to this shape. Although a bat's wing is essentially a sheet of stretched skin, its attachments to the body, legs and wing bones give it the necessary curvature. The fact that the aerofoil is 'hollow' on the underside is not a problem, since the air moves around it as it would a more solid structure. Because the bat can move its legs, fingers and thumbs it can change the cross section of the aerofoil, an important ability, as we shall see.

As the wing travels forward, the air passing over it is deflected downwards due to its aerofoil shape. Since Newton's third law (every action must have an equal and opposite reaction) must be obeyed, there must be an equal and opposite force on the wing itself, pushing it upwards. If the wing is moving horizontally, most of the force acts vertically: this is the *lift*. Because the wing meets some resistance, or *drag*, from the air, there is a tendency for the wing to be pushed backwards. The net direction of the forces acting on the wing is therefore up and back as shown in Figure 3.2.

The animal can only go forward if something else pushes it forward: it needs a *thrust* component. In aeroplanes the job of developing thrust is done by the

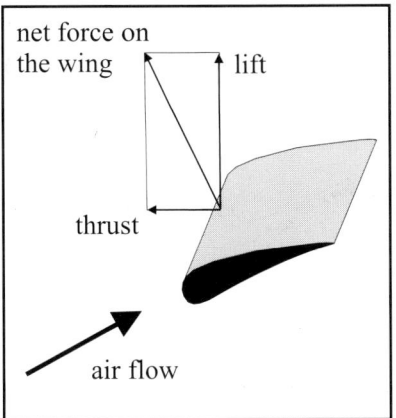

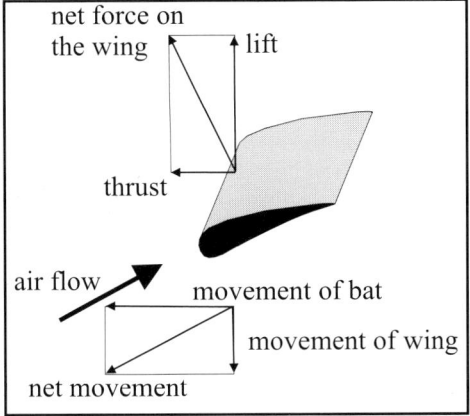

Fig. 3.3 How a gliding bat generates lift and thrust.

Fig. 3.4 How a flying bat generates lift and thrust.

engines. A gliding animal tilts its body and wings down at the front to go forward (Fig. 3.3). The wing now generates lift and thrust, but there is an obvious cost: the animal glides down towards the ground; it cannot glide horizontally.

Powered flight requires a more complex solution. The wings must be moved up and down so that the combined movement of the whole bat and its wings generates lift and thrust. For a bat going forward horizontally, with tilted wings beating down, the resultant air flow gives the desired effect (Fig. 3.4). The power for forward flight now comes from the massive flight muscles that pull the wing down.

The final complexity comes in the upstroke. If the downstroke produces lift and thrust, then the upstroke will produce lift and drag. Unless the wing changes orientation or shape during the upstroke the drag will cancel the thrust and the bat will not move forward at all. This is why the wingbeat of bats (and other flying animals) is asymmetrical. On the downstroke the wing has an appropriate angle of attack to the air flow to generate lift and thrust. On the upstroke the angle of attack may be reduced to zero, that is the wing is 'feathered' so that it travels in the plane of the air flow without generating any significant forces. Alternatively, or in addition, the wing may be partially folded to reduce drag. Both are often used together (Fig. 3.5).

The machinery of flight

The wing

A bat's wing is a double layer of skin supported by the extended bones of the forelimb, the hind limb and, in all British bats, the tail. Although greatly extended, the bones of the wing can be easily identified and matched with those of other mammals (Fig. 3.6). The leading edge of the wing is supported by the humerus, radius and the second and third digits. The first digit, the thumb, is free and is used for climbing and grooming. The fourth and fifth digits extend backwards to support the trailing edge of the wing, aided by the hind limb. The tail also supports part of the wing. The second and third fingers are joined at their tips by a ligament to strengthen the leading edge. Most

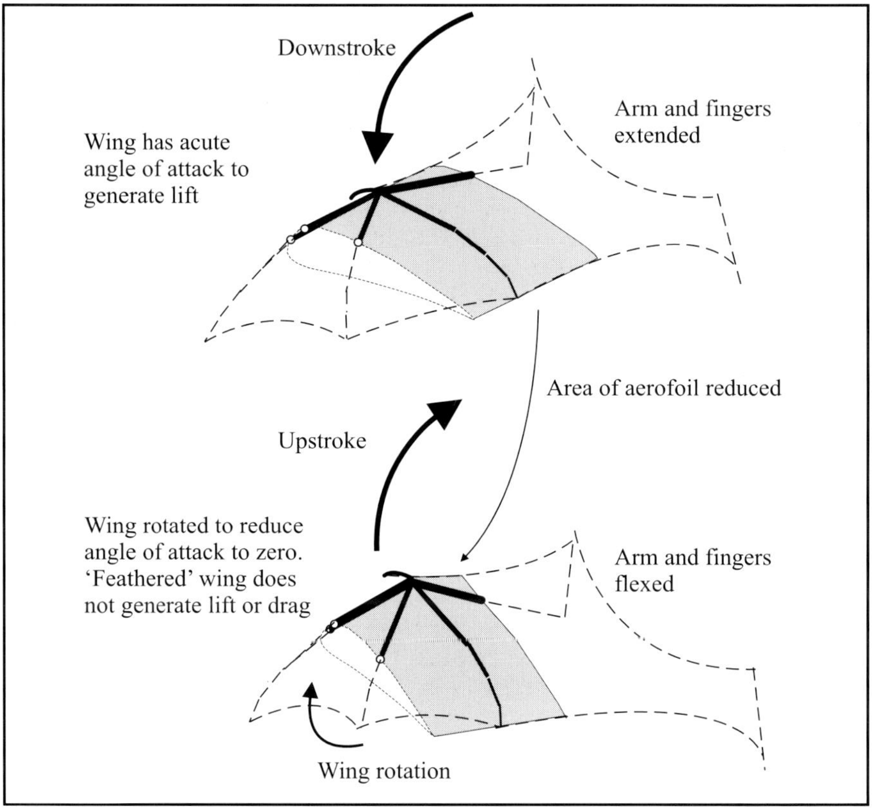

Downstroke

Wing has acute
angle of attack to
generate lift

Arm and fingers
extended

Area of aerofoil reduced

Upstroke

Wing rotated to reduce
angle of attack to zero.
'Feathered' wing does
not generate lift or drag

Arm and fingers
flexed

Wing rotation

Fig. 3.5 Wing beat asymmetry: how bats fly forward.

of the extension of the 'fingers' is actually due to elongation of the bones of the wrist, the metacarpals, not the finger bones or phalanges.

Bats have evolved a very elegant way of increasing the efficiency of the wing skeleton during flight by rigidly locking together the humerus and the shoulder blade (scapula) during the downstroke. All the bones are light but strong. To achieve this, each of the major stress-bearing bones has its greatest diameter in the plane subjected to the greatest bending forces during flight: bone is not laid down where it is not needed.

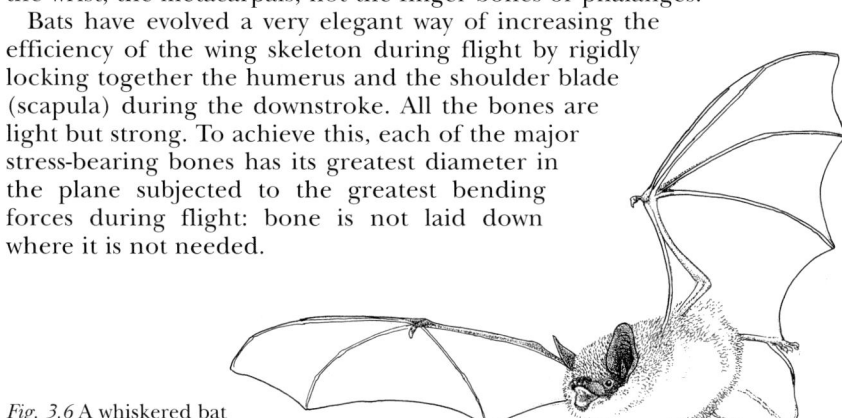

Fig. 3.6 A whiskered bat in flight, emphasising wing features.

Adaptations for flight extend to other parts of the skeleton. Bats, like birds, have relatively short and streamlined bodies and many of the bones of the vertebral column are fused to increase the rigidity of the thorax. A network of fibres made of 'stretchy' elastin and inextensible but flexible collagen control the shape of the wing membrane in flight. Without them the thin wing would bulge out due to the powerful aerodynamic forces acting on it. The membrane also has some small muscles to control folding, and a rich nerve and blood supply: a damaged wing membrane rapidly repairs itself. Each part of the wing membrane has a slightly different function. The armwing or plagiopatagium (Fig. 3.6) lies between the body and the fifth finger. It generates most of the lift during flight. The handwing or dactylopatagium between fingers three and five generates most of the forward thrust. The uropatagium lies between the two hind limbs and is supported by the tail. It generates lift and may be used as an air brake. Many bats use it to catch insects too. In front of the humerus and radius, on the leading edge, is a small area of the wing known as the propatagium. It generally tilts forward and down, relative to the rest of the wing, increasing the camber and helping to give the wing its aerofoil shape. In slow-flying, manoeuvrable bats the propatagium may be quite large and its angle controlled by the thumb. Just like an aeroplane wing flap, it can be lowered, increasing the camber to increase the lift at slow flight speeds, which prevents stalling. It also increases drag, but this is not so important for slow-flying bats. Another difference between fast- and slow-flying bats is in the shape of many of the wing bones. Fast-flying bats have bones with flattened profiles to make them as streamlined as possible. Some slow-flying bats have bones with rounded profiles, so that they stand proud of the wing surface. This may be another adaptation for slow flight. As flight speed decreases, lift decreases and the bat may compensate by increasing the angle of attack. If the angle of attack is too great, air flow over the wing becomes turbulent ('breakaway') and the wing stalls with complete loss of lift. Bats appear to do what aircraft engineers have only recently learned how to do: they generate controlled microturbulence close to the wing surface to prevent breakaway. The microturbulence is produced by the raised wing bones or by hairs on the leading edge: in aircraft it is done by adding corrugations or arrays of bumps to the wing surface. Whatever the technology, the aim is to carefully mix the air close to the surface of the wing, reducing the velocity gradient (the rate at at which air speed changes close to the wing) in this critical region, since turbulence occurs when this gradient exceeds a critical value.

The flight muscles

Birds have evolved a very straightforward power source for flight. Attached to the keel, the very obvious projection from the breast bone (sternum), is a huge pair of pectoralis muscles that effectively provide all the flight power. Bats have evolved very differently, perhaps because their forelimbs are still important for walking and climbing as well as for flight. Five muscles provide power for the downstroke and two muscles assist the upstroke in bats. There is no bony keel on the sternum, but some bats do have a sheet of ligament to which the flight muscles attach.

The muscles must be able to work for very long periods without fatigue. This 'aerobic exercise' demands the use of oxidative metabolism: fuels are used up in reactions that rely on a constant supply of oxygen from the blood. The muscles therefore have a very rich blood supply, to provide the metabolites to be broken down, oxygen to fuel the breakdown, and a means of removing the waste products of metabolism. The muscle cells are rich in the oxygen-carrying protein myoglobin, which strips the oxygen from the haemoglobin in the blood, making it available for aerobic metabolism in the muscle.

The circulatory and respiratory systems

The oxygen needed to power flight arrives in the lungs and is taken to the muscles in the blood by the circulatory or cardiovascular system. The cardiovascular system in bats has improved performance relative to other mammals, but the respiratory system appears to be unchanged. Although the heart rate of a resting bat is similar to that of terrestrial mammals of the same size, in flight it may increase as much as two- to six-fold. Only in birds is a comparable increase seen. A small bat such as a pipistrelle typically has a heart rate of 200–450 beats per minute at rest, rising to 800–1,000 in flight. Bats also have hearts two to three times bigger than those of other mammals of similar size. Together these two factors greatly increase the rate of blood delivery to the flight muscles: more blood is pumped every stroke, and there are more strokes per minute. The oxygen collected in the lungs is carried to the muscles bound to haemoglobin in the red blood cells. Those bats studied have more red blood cells than other mammals and even birds, and can therefore carry more oxygen. Bats achieve their high performance by fine-tuning typical mammalian physiology, rather than by using novel systems.

Echolocation

Echolocation is a highly sophisticated sonar system. Bats emit short, high frequency bursts of sound and use the information carried in the echoes to construct a sound 'image' of their environment. This constantly updated image contains the information most vital to them. They need to know the precise positions in space of objects around them, and the nature of these objects. A bat needs to know its own speed and direction of travel relative to these objects. It needs to know of the presence of insects and their location, speed, direction of travel and perhaps even their identity. This section describes what is known of how bats achieve all of this. Inevitably, the description makes the bat sound like a modern fighter pilot wired to the onboard computer. But we must not be misled. If I were to describe how we use vision to see our complex world it would be equally impressive, yet we take it for granted. The human brain works simultaneously and rapidly on a bewildering range of tasks, and we perform our daily deeds without a thought to this complexity. To a flying bat, echolocation is just another component of the sensory and motor (movement) tasks it has to perform. Most of the processing is carried out rapidly and automatically. In trying to study and understand it all we marvel at the processes involved and the tasks performed. Yet at the same time these very tasks and processes must be simplified and underestimated before we can even begin to study and understand them. This section describes some of the 'simple' tasks that have been investigated, but also hints at some of the complexity that researchers are now beginning to consider: the subject of echolocation is becoming more and more interesting.

Making and hearing sounds

Bats make and hear sounds just like most other mammals: they generate the sound (echolocation pulses) in the larynx (voice box) and listen with their external ears. This is worth stating because dolphins and whales, the only other mammals to have a sophisticated sonar system, make their sonar clicks in their nasal passages and listen through an acoustic 'window' in their jaw. The larynx of bats is large relative to that of other mammals, since echolocation calls need to carry a great deal of energy. Air is passed across the vocal chords of the larynx to vibrate them. The tension in the chords determines their vibration frequency and hence the frequency of the emitted sound. This tension and hence the sound frequency is controlled by muscles in the larynx.

Most vesper bats (family Vespertilionidae) emit their echolocation sounds through the mouth, which is why virtually all photographs of flying bats show them with their mouths open. Horseshoe bats emit their calls through the nostrils, which explains why they have complex noseleaves (projections of skin and cartilage above their nostrils). In those species studied, the noseleaf focuses the sound into a narrow beam in front of the bat. By analogy, our visual system works best for objects in front of us. Although we do have peripheral vision, if we want to see an object to one side clearly we turn our heads. This is particularly important to bats, since they have to expend energy sending out echolocation calls, and it would waste energy to send sounds to the periphery. By contrast, vision gathers information using the free energy of the sun's light and comes at little cost.

Despite the high intensity of echolocation calls the returning echoes are very weak (p. 40), so bats have evolved various ways to detect the echoes. One way is to have relatively large external ears or pinnae (Fig. 3.7). Different bats use echolocation in different ways and the size and shape of the pinna is often a good guide to a bat's way of hunting for food: bats that rely on detecting weak sounds have large external ears to gather that sound. The shape of the ears of horseshoe bats makes them particularly sensitive to the frequencies of their constant frequency calls. Because bats listen in stereo, they can tell where a sound is coming from, and the shape of the ears of all bats studied has been shown to enhance this ability. A cartilaginous structure called the tragus sits inside the pinna (Fig. 3.7). It projects from the base of the ear and masks to some extent the direct entry of sound to the middle ear. Most bats hear best

Fig. 3.7 Variation in ear and tragus size and shape: 55kHz pipistrelle and Natterer's bat.

over an angle of 30–40° directly in front of them and the combination of pinna and tragus shape determines this. Once inside the ear sound travels to the eardrum and causes it to vibrate. This vibrates the tiny ear ossicles (a system of bone levers between the two eardrums) that amplify the vibrations en route to a second eardrum, the oval window. The oval window connects to the inner ear, and the first part of the auditory system (the cochlea) that begins the complex process of sorting and interpreting the echoes. The cochlea is a conical, fluid filled spiral. At the bottom of the spiral are sensory cells that respond to high frequency sounds and at the top are cells sensitive to low frequency sounds. Each sensory cell in the cochlea connects to a different set of cells in the auditory centre of the brain. The brain knows the frequency of the sounds detected, by which cells receive nervous signals, since each codes for a different frequency.

The structure of echolocation calls

Bat echolocation calls come in a wide range of forms, but they have several well-defined features. The following are some of the basic properties of the calls.

Frequency (pitch)

Most bats use calls of very high frequency (high pitch). In the vast majority, most of the energy of the call is in the range 20–80 kHz (kHz or kilohertz is thousands of cycles of air vibrations per second). Human hearing rarely extends beyond 18 kHz so we cannot hear most bats and the sounds are referred to as ultrasounds. Bats need to use high frequencies because small objects, such as insects, can only be detected using high frequency sounds.

Energy (intensity or loudness)

Sound looses its energy as it travels through the air: the air itself absorbs the sound, a process called attenuation. The higher the frequency the greater the attenuation, so ultrasound calls lose their energy very quickly. Bats therefore need to use very high energy calls if they want them to travel very far. The intensity of a pipistrelle's call, measured 10 centimetres in front of it, is as much as 120 decibels: that is the equivalent of holding a domestic smoke alarm to your ear. Despite this, the echolocation systems of most bats operate best over ranges of just 1–10 metres. This is because another process, in addition to attenuation, reduces the amount of sound energy reaching a particular point. If a bat sent out a pencil thin beam of sound, and there was no attenuation, then the amount of sound energy bouncing off a moth 1 metre away would be much the same as that from a moth 5 metres away. But as we know, bats send out a broad beam of sound, and, as it travels, its energy is spread over a greater and greater area. The energy at a particular point therefore falls, by a rule called the inverse square law (the energy is inversely proportional to the square of the distance travelled). So, the intensity of a sound wave hitting a distant moth is much lower than that hitting a nearby moth. The echo that the bat must detect is therefore weaker. Since the returning echo is also subject to attenuation and the inverse square law, the bat hears a *very* weak echo. Perhaps one of the most impressive adaptations of bats is their ability to detect these very weak sounds.

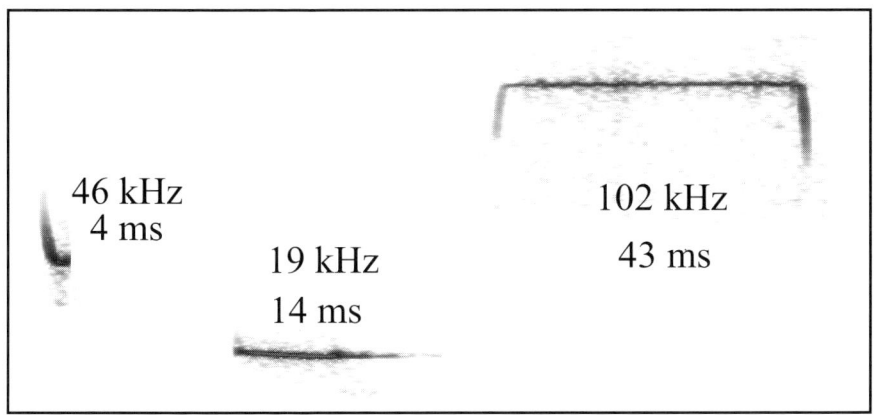

46 kHz
4 ms

19 kHz
14 ms

102 kHz
43 ms

Fig. 3.8 Echolocation calls of British bats to show variation in call duration. From left to right, pipistrelle, noctule and lesser horseshoe bat.

Duration

Most bats use very short calls (Fig. 3.8). A pipistrelle call is usually less than 5 milliseconds (thousandths of a second) long. A noctule call may be up to 20 milliseconds long and a horseshoe bat call is frequently as long as 50 milliseconds. In most cases, echolocation involves the emission of a call, reception of the echo, interpretation of that echo and then emission of the next call. Since sound travels at 340 ms^{-1} (metres per second), it takes less than 6 milliseconds for a bat to receive an echo from an insect 1 metre away. If it used a pulse longer than 6 milliseconds, it would be listening to the echo before it had finished sending out the pulse. Overlap between call and echo impairs the performance of the echolocation system of most bats, so pulses are usually made short enough to avoid it. As a bat detects and approaches its target, the time between pulse and echo becomes shorter. This enables the bat to update its information about its prey more frequently, but it may have to shorten its pulses as it does so. A typical British bat emits pulses at about 10–12 Hz, that is every

Fig. 3.9 Feeding buzz of Daubenton's bat.

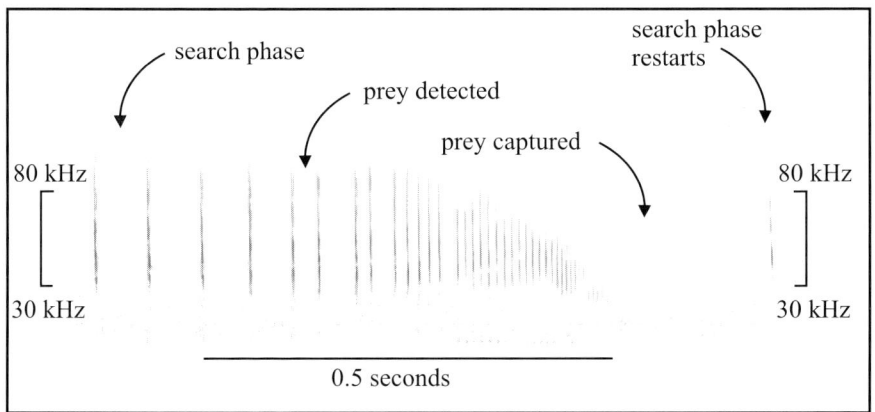

search phase

prey detected

prey captured

search phase
restarts

80 kHz

30 kHz

80 kHz

30 kHz

0.5 seconds

100 milliseconds or so: one pulse in every wingbeat. During the final capture phase of a 'feeding buzz' (Fig. 3.9) this may increase to more than 150 Hz, and call duration may fall from, for example, 5 milliseconds to less than 0.5 milliseconds.

Structure

Some bats use frequency modulated (FM) calls that sweep down from a high to a low frequency, others keep the frequency of their calls almost constant throughout (CF calls), and others use a combination of FM and CF components (Fig. 3.8). The average frequency of calls within the ultrasound range can vary from less than 20 kHz to over 100 kHz in British bats. Good reasons can be given for all these differences, but it is not always easy to say why a particular bat uses one call structure rather than another. It is better to show how bats use echolocation to perform essential tasks and in the process show how call structure can enhance performance of these tasks.

Things bats needs to know, and how they find out

As a bat flies through its favoured foraging site looking for food, it has to perform many tasks simultaneously. Breaking the processes down into simple questions enables us to see how echolocation is, or might be used, to answer these questions. For the moment, assume the bat is concentrating only on its food and that it has detected potential food by the general cues of size, shape and pattern of movement. Bats do have specific means of detecting flying insects, but these are best dealt with after the basics (p. 45).

How far away is the insect?

This is the easiest question to answer. Since sound always travels at a constant speed in air, the bat can 'calculate' distance between itself and its prey from the time that elapses between emitting a pulse and receiving the echo. These times are short: a Daubenton's bat is unlikely to detect a small insect much more than 3 metres away, so the distance travelled by pulse and echo is 6 metres. The total time elapsing between pulse emission and echo reception is thus less than 18 milliseconds: about one-sixtieth of a second. How accurately can these short distances be measured? This depends upon the bat and the type of echolocation call it uses. Bats that use high frequency FM calls, such as Daubenton's bats and other *Myotis* species, can resolve distance down to about 5 millimetres when given tasks to perform in the laboratory. Horseshoe bats using predominantly CF calls can do little better than 12 millimetres. A bat that determines distances to 5 millimetres is resolving time differences in echo arrival down to a mere 20 μs (millionths of a second).

How big is it?

The bigger the insect, the more intense the echo the bat receives. Of course a small insect near the bat may give an echo of similar intensity to that from a large insect some distance away. The bat therefore needs to combine information on target distance and echo intensity.

In which direction does it lie?

As discussed earlier, the pinna and tragus give some directionality to sound reception, but bats rely primarily on the fact that they have two ears to answer

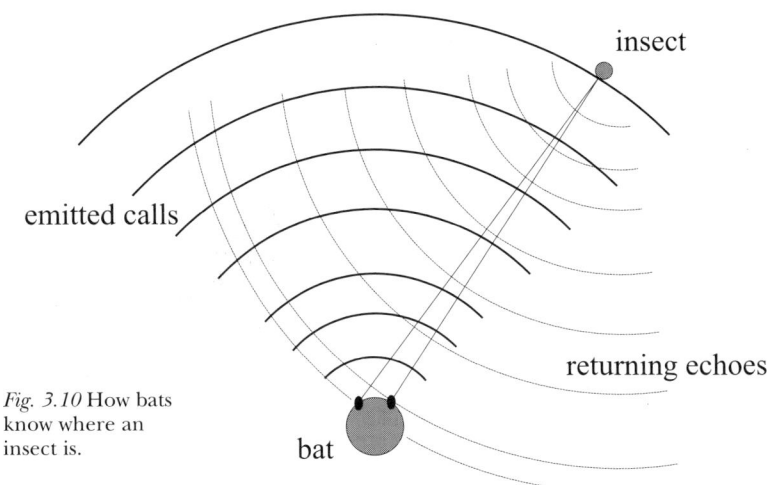

insect

emitted calls

returning echoes

Fig. 3.10 How bats know where an insect is.

bat

this question. If, for example, a potential meal is flying to the right of the bat, an echo reaching the bat from this insect reaches the right ear a little sooner than it reaches the left ear (Fig. 3.10). Since the left ear may also be partly in the shadow of the head, the echo may be less intense. The bat may use both differences in the arrival time and intensity of echoes at the two ears to determine where an insect is in the horizontal plane. Arrival time differences are only 1–2 milliseconds.

The big brown bat, *Eptesicus fuscus*, a relative of our serotine, resolves horizontal angles to 1.5°. Locating the source of a sound in the vertical plane is a less intuitively obvious problem. However, it appears to involve the use of the tragus. The big brown bat resolves vertical angles to 3.5°, but when the tragus is gently held down, this resolution is reduced to about 7°. Vertical angles are probably determined by analysis of the complex multiple reflections that travel down the external ear to the eardrum. However it is done, this degree of accuracy is quite remarkable.

How fast is it flying and in what direction?

Having determined in which direction and how far away an insect is, using a single pulse, a bat can repeat the calculations from pulse to pulse. Does it simply plot a new position, or does it combine information from several pulses to calculate speed and direction of travel? It seems that some bats can do the latter, since they are able to fly on interception courses. An accurate interception course can only be flown if the bat knows where the insect will be at a given moment in time, and for this it needs to know speed and direction of travel. A bat that only knows where an insect is at a given moment must necessarily follow it and cannot fly an interception course.

What is it?

Many bats, much of the time, do not appear to select their prey, but eat whatever is available to them. If a range of different insects is available then they are eaten in proportion to their abundance. Some selection may occur simply because a small bat may be physically unable to capture and eat a large moth, and the echolocation calls of some bats may not be able to detect small insects. Even if a bat is capable of recognising different insects, whether or not it chooses to be selective will depend upon which strategy will benefit it most. In most situations a bat will aim to maximise its energy intake whilst minimising the energy it expends foraging. Selecting particular insects may only be beneficial if they are particularly abundant, nutritious or easy to catch. In a nutshell, the ability to recognise different insects is not the only factor determining how bats feed. This section is concerned with whether or not bats are able to tell different insects apart: optimal foraging strategies will be discussed later (p. 84).

The best experiments to address this question were carried out on greater horseshoe bats, *Rhinolophus ferrumequinum*. As is the case with many bats, they can be brought into captivity for a short period and taught to perform tasks. In this case, the bats were shown different large insects, such as craneflies, moths and beetles. The bats would echolocate to examine each insect, and microphones placed next to the stationary bats recorded the echoes. The experimenters were therefore able to build up a library of echoes: the bats' perception of their prey. These echoes were then used in the experiments rather than real insects, acting as 'phantom' targets that could be subtly changed to investigate the bats' perceptive abilities. What information might these echoes contain? Greater horseshoe bats emit FM/CF/FM calls at frequencies of about 83 kHz. There is a very short rising FM component, a long CF phase, and then a short falling FM phase. Ignoring the brief FM components for a moment, consider the call as pure CF. If the call bounces off a flat, stationary target, the bat hears an echo essentially the same as the emitted pulse. But what if the target is a complex, fluttering insect? As the insect's wings beat up and down they rhythmically show different surfaces to the bat, from the flat face of the wing to the wing edge. The intensity of the echo therefore changes rhythmically. If the bat can detect this 'amplitude modulation' in the echo it can theoretically calculate the insect's wingbeat frequency, which could give it clues about the insect's size and identity.

In the experiments, to eliminate this variable, the biologists electronically manipulated the echoes so that all the 'phantom' insects had the same wingbeat frequency. A moving target provides additional information. Consider how the siren on an ambulance or fire engine changes pitch as it passes. As the vehicle approaches the siren sounds high pitched, because the sound waves from it are squeezed together. As it passes, the pitch is suddenly lowered, because the sound waves are now stretched. This is known as Doppler shift. As an insect beats its wings up and down, they alternately move towards and away from the bat, and the speed at which they do this determines how much the frequency of the echo is changed. A horseshoe bat emitting a CF call hears an echo that oscillates above and below the call frequency. This is known as frequency modulation. Different insects were shown to have distinctive echoes due to complex amplitude and frequency modulation of their calls. To make matters more complicated, each insect produces a spectrum of echoes as it is rotated in front of the bat.

The experiments were as follows. A bat echolocated towards two targets and received 'phantom' echoes from each. It was rewarded with a juicy mealworm if it could identify the echo from, for example, a crane fly. Once all the bats could do this, they were 'asked' to make the same choice, but this time the crane fly echo was from an insect with a different orientation relative to the bat, for example, one flying away from it rather than across its path. Could the bats still identify the crane fly? Indeed they could. Their ability to recognise echoes from different insects did not depend upon the insects always facing the same direction. Which components of the echoes were important to the bats, the amplitude modulations or the frequency modulations? The answer was both, but interestingly some bats relied more on one type of modulation than the other. Horseshoe bats are ideal for such studies: the relative simplicity of their calls makes the task of finding out how they work easier for the biologists studying them. There is no reason to suppose that bats with more complex calls cannot perform similar tasks.

To what extent do bats make use of these abilities in the more complex environment of the real world? Since they clearly have these abilities, it would be surprising if they did not use them, and the next chapter considers evidence that suggests that horseshoe bats, and some other species, do indeed select their prey.

Specific prey detection methods and why echolocation calls vary

The frequency and amplitude modulation of calls described above can be used as a simple means of prey detection. Any bat that echolocates using relatively long calls can have a 'flutter detector'. The echoes produced by flying insects will be distinctive and attention-grabbing: consider how much more effective flashing cycle lights are than old-fashioned lamps with a steady beam. If a call is 50 milliseconds long, then within one echo a bat can record at least one full wingbeat of any insect beating its wings at more than 20 Hz. A noctule using a 20-millisecond call will detect the full wingbeat of insects beating their wings at more than 50 Hz. Most small insects beat their wings at higher frequencies, and flutter detection may not even require the reception of a full wingbeat. These effects may be visualised most easily on CF calls, but FM calls will be modulated in the same way. Certainly many bats using long, 'narrowband' FM calls may take advantage of flutter detection.

As frequency increases the attenuation of sound in air increases. Under ideal conditions a 100 kHz call may have a range of up to 10 metres, but a call of 30 kHz may be useful over as much as 30 metres. A noctule call at 18 kHz will therefore travel further than the 105 kHz call of a lesser horseshoe bat. One way to further increase the useful echolocation range is to concentrate the energy into a 'narrowband' call: one with a narrow bandwidth, or a small range of frequencies. The bat only hears the echo if sensory cells in its hearing system are activated. Each sensory cell is fired by sound of a particular frequency and requires sound of a certain threshold intensity to activate it and send a signal to the brain. Since the echo can carry only so much energy, concentrating it into a narrowband call increases the energy in each frequency and therefore increases the chance of activating the hearing system.

All bats can hear over a wide range of frequencies, and are particularly sensitive to high frequency sounds, since these are used not only for echolocation, but also for much of their social repertoire. Social calls may have components lower in frequency than echolocation calls, and may even be audible to humans, but they are still relatively high frequency. In addition, the hearing of many bats is particularly sensitive over the frequency range used in their echolocation calls, increasing their ability to pick up weak echoes. This increase in sensitivity is most marked in bats that use narrowband FM or CF calls. They have what is known as an acoustic fovea: over a very narrow frequency range, hearing sensitivity may be increased by several orders of magnitude. In fact, this zone of enhanced sensitivity is typically at slightly higher frequencies that those in the echolocation call. This compensates for the fact that the motion of the bat increases the frequency of the echoes through Doppler shift.

If CF or narrowband calls have such clearly advantageous properties, why do most bats use FM calls? A discussed earlier, bats needed to use high frequency calls to detect small insects. Whether an insect reflects or refracts sound depends upon the relationship between the size of the insect and the frequency of the sound. To be reflected, rather than 'bent around' the insect, the sound must have a wavelength similar in magnitude to the size of the insect. Small insects demand short wavelengths, which means high frequency. The same applies to details of the structure of insect prey or the environment. Sound reflecting from complex surfaces will produce interference patterns, resulting in echoes that will convey information about the surface. If a target is structurally complex, more information about it will be carried in echoes from complex FM calls than in echoes from CF calls. Furthermore, because FM calls typically start at high frequencies and sweep low, the bat 'knows' the order in which it will receive the echoes from these components. Whatever complex modulation occurs in the echo, the bat knows which component of the echo to relate to each stage of the emitted pulse.

So far only the insect the bat is chasing has been considered. All the subtle information a bat receives about its environment comes from the modulation of its echolocation calls caused by that environment. Most bats hunt their prey in a complex and changing habitat. To learn more about this complex environment, the bats need to use more complex calls.

In fact, the calls of most bats are more complex than simple FM or CF. Many FM bats, such as the pipistrelles, terminate their calls with a narrowband tail, and the calls of CF bats usually have FM components at the beginning and end (Fig. 3.8). Horseshoe bats typically shorten the CF component and add a very broadband FM tail as they approach a target. Some bats, such as the noctule, alternate narrow and broadband FM calls (3.11). Many bats use harmonics, multiples of the 'fundamental' frequencies (the lowest frequency component of the call) (Fig. 3.11) to increase the bandwidth and complexity of their calls.

Since vision is such an important sense in humans, analogies using vision can help us to understand how bats use sound. Movement (Doppler shift)-induced changes in returning echoes may be thought of as colour changes. A change in the frequency of sound changes pitch, and a change in the frequency of light changes colour. These 'spectral' changes may be used by FM bats to detect insects in the complex habitat of a wood. As leaves and branches move in the wind they will produce 'colour' changes that the bat recognises. The movement of an insect, either in flight or walking on a leaf may produce other, equally characteristic changes. It is not necessary for the bat to resolve all the details of its environment: detection of these brief beacons of 'colour' may be the key to successful prey detection. In theory, a bat echolocating at 20 kHz can detect an insect shifting its position on a leaf by less than a millimetre. Daubenton's bats may detect an insect by a characteristic 'colour' change against the changing and equally characteristic 'colour' surface of the moving

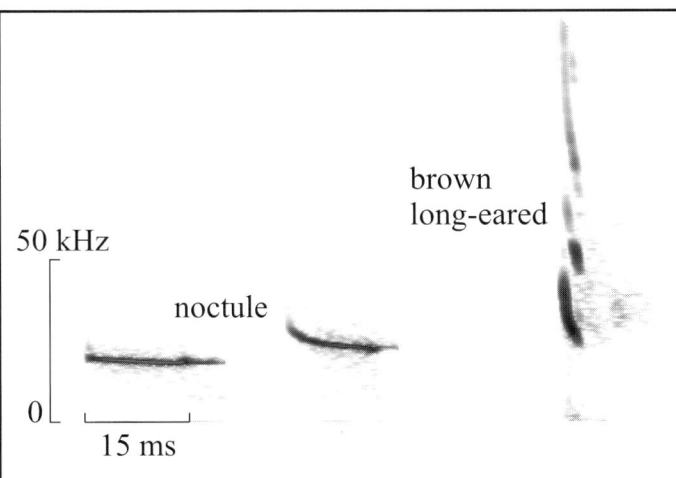

Fig. 3.11 Some complex bat calls: the two calls of the noctule, and the multi-harmonic call of the brown long-eared bat.

water. Many gleaning bats hover with their heads motionless. This may be an adaptation to eliminate some of the complex spectral changes due to their own motion.

Doppler shift could be used as a means of calculating the speed and direction of prey from a single echo, rather than the method described above, which depends upon the integration of information over several pulses. However, all the species investigated to date use Doppler shift to determine their own speed relative to their environment, and pulse to pulse information of position to monitor the movement of prey. This use of different techniques to monitor different components of sensory input will almost certainly prove to be an even more subtle and fascinating topic for future research.

Prey capture

The time interval between a bat first detecting an insect and capturing it is usually less than a second, but in this time several things happen. After an insect is detected, the rate of emission of echolocation pulses increases from around 10 Hz to up to 150 Hz just before capture (Fig. 3.9). As the bat approaches an insect the pulse-echo delay time decreases, so the bat is able to send out pulses more frequently. This is not only inevitable, but also advantageous, since the bat can update its information about the nature and whereabouts of its target more frequently. To avoid pulse-echo overlap call duration is shortened to as little as 0.2 milliseconds. During the approach phase an FM bat may increase the bandwidth of its call by increasing the bandwidth of the fundamental, by adding harmonics, or both. This can improve the accuracy with which it measures the distance to its target and its flight path. It can also potentially provide more information about the identity of the target. During the final phases of capture, bandwidth may be decreased and call frequency may decrease significantly. Just why this is done is not known. Call intensity is also lowered during the approach, to counteract the increased intensity of the echo from a nearer target. This keeps the intensity of the returning echo at a level best suited to the hearing of the bat.

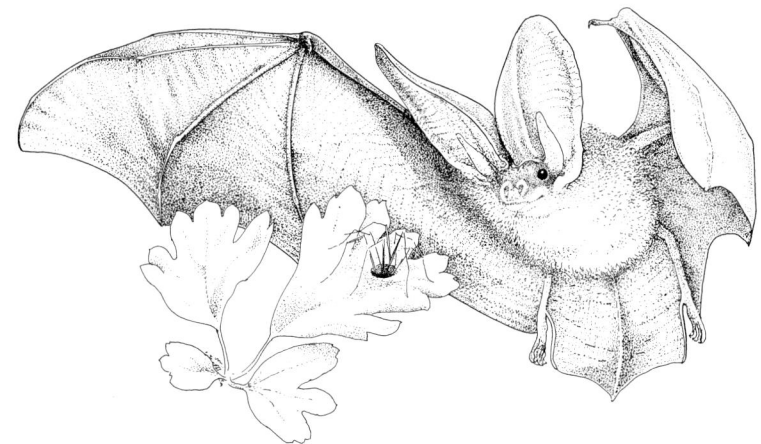

Fig. 3.12 A long-eared bat gleaning a harvestman from a leaf.

Passive hearing and vision

Although echolocation is the primary means by which temperate bats find
their way through their environment and catch food, some bats make use of
other methods. Several gleaning bats, that capture non-flying insects off the
ground or trees for example, may 'switch off' their echolocation as they
approach their prey (Fig. 3.12). They then rely on passive hearing to catch
their food, listening for sounds generated by the prey flapping their wings,
moving or calling. Long-eared bats certainly use this method. This subject will
be discussed in more detail in the next chapter (p. 79).

   Bats have other senses that may be useful for feeding, but in British bats only
vision is likely to be important. The eyes of bats are adapted for night vision,
sharing many features with those of other nocturnal mammals. There is some
evidence to suggest that their night vision is often good and may be used for
navigation and even prey capture. Vision is most definitely used in prey capture by some gleaning bats, such as the California leaf-nosed bat, *Macrotus californicus*. Large eyes are a notable feature of long-eared bats, and I would be
surprised if they did not use vision to find their food.

Torpor and hibernation

Bats evolved in the tropics, and without the ability to use torpor and to hibernate, it is unlikely that they would ever have invaded temperate parts of the
world, since food is scarce or absent for much of the year. Although some bats
are migratory, few temperate bats migrate far enough to feed throughout the
year and most migrate only to find suitable hibernation sites. Throughout
their lives, insectivorous bats must consume sufficient food on a daily basis to
provide the energy for flight, mating, reproduction, and simply maintaining
their body temperature. Much of the time temperate bats are homeothermic
– like humans they maintain their bodies at a constant temperature, significantly above that experienced in the typical British climate. They maintain this
constant, high body temperature by burning stored fuels in the form of fats

and some carbohydrates. Because they are small, with proportionally small volumes to generate heat and a large surface over which to lose it, they have to generate heat at a high rate. Since they are small and because they fly, they cannot carry large fuel reserves as these would make flying energetically expensive. This is a demanding combination: bats really do live life on a knife edge. The insects that bats feed on are often low in energy and sometimes hard to find, being patchily distributed or subject to changes in weather conditions. This further stacks the odds against the bats.

Daily torpor and seasonal hibernation are key components of the survival strategy of temperate bats. Torpor (a controlled lowering of body temperature) helps bats to balance their delicate energy budget. Without torpor, almost every day would be a battle to eat enough to sustain their energetic way of life. The evolution of torpor in temperate bats has increased their options and their chances of survival in marginal habitats and climates. Bats have the ability to switch from homeothermy to heterothermy. A heterotherm can reduce its body temperature and maintain it within set limits. The temperature drop may be just a few degrees or it may be very substantial. An active homeothermic bat has a core temperature of about 38–40°C. The same heterothermic bat may reduce this to 15°C, the temperature of its tree roost in summer, or 2°C, the temperature of its hibernation site in a cave. By reducing its body temperature it reduces the rate of all the body's chemical reactions, and thus reduces the rate at which stored fuels are metabolised. A small bat such as a whiskered bat, *Myotis mystacinus*, hibernating at 2°C, may use energy at a rate less than 1 per cent of that of a homeothermic bat at rest. Even in the summer, the same bat can save 15 per cent of its daily energy requirements by lowering its body temperature to 20°C in its day roost.

What is torpor?

Reduction of body temperature in heterothermy is not a straightforward process. As the temperature falls and the body's metabolic processes slow down, steps must be taken to keep essential life-support systems running at 'maintenance' levels. As the bat relaxes control of its core temperature and cools, breathing rate slows, heart rate slows and the oxygen supply to the tissues therefore falls. Blood vessels to the limbs and to many organs and tissues shut down to slow or even halt the flow of blood. Blood volume decreases by the loss of plasma, and excess red blood cells are stored in the spleen. The most important organs, such as the brain and heart, still have work to do and receive a reduced but regular supply of blood. In deep torpor a bat may breath as infrequently as once every hour or so and its heart rate will be 10–60 beats per minute (compared to 900–1,000 in flight). Daily torpor in summer is rarely this deep, but summer torpor and winter hibernation are physiologically similar. They differ only in the depth and duration of the torpid state. In deep torpor small fat stores can go a long way. By using just a few milligrams of fat each day, a well-fed bat may have the potential to stay in hibernation for over a year. However, even under ideal conditions, bats arouse every few days or weeks, even if they do not fly or leave the roost. Arousal may take just 30 minutes and bats may quickly return to torpor, but the energy used is substantial. Frequent arousal therefore reduces the time that bats can stay in hibernation without feeding. Bats may arouse for a number of obvious reasons such as feeding and drinking, as will be discussed later (p. 53), but there are many occasions when

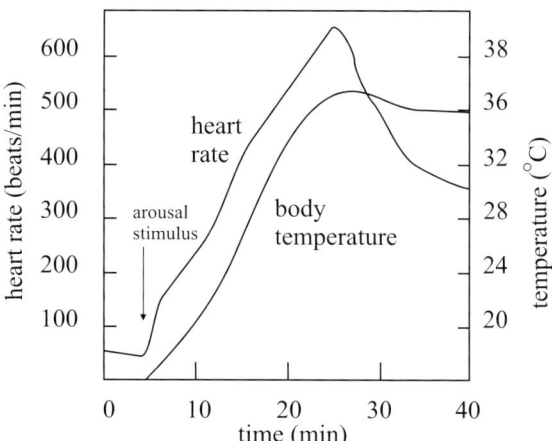

Fig. 3.13 Physiological changes that occur during arousal from torpor.

the reasons are not obvious. Given the large energetic cost, arousal must serve important functions, even if these are not yet understood. It may be that it is necessary for physiological adjustments or for immunological reasons, to prevent or fight disease.

To arouse, a bat first increases its heart rate and breathing rate (Fig. 3.13). Blood is sent to the brown adipose tissue, a special deposit of fat cells on the back. These cells are capable of generating large amounts of metabolic heat to warm the blood, which then goes on to warm the rest of the body. When the muscles are warm enough to work, they begin to shiver, producing more heat, and the bat rapidly increases its body temperature to its resting homeothermic level.

How do bats use torpor?

When weather conditions are marginal and the food supply is unreliable, a bat can choose not to feed, but to remain in the roost at night and become torpid, feeding only on the best nights. Even when the food supply is good, it may be beneficial to feed and then enter torpor on return to the roost, reducing the rate at which this hard-earned energy is used. Whether or not a bat can make use of torpor depends upon the species, the physiological state of the individual and the ecological and behavioural demands of the moment. In the early spring, both sexes of most species will make regular use of daily torpor. As summer approaches, however, the females become pregnant. To maximise the rate at which the foetus develops, females will remain homeothermic whenever possible. When food is in very short supply, a female may lower her body temperature to save energy, but at the consequence of slower foetal growth. Even after birth the need to produce milk for their offspring ensures that females remain homeothermic much of the time (p. 62). Males have none of these commitments and are able to spend much more time in torpor than females, although during the mating season, males of some species actively advertise for females or guard those they have attracted. They must therefore remain homeothermic for some of the time they are not feeding. As winter approaches, both sexes spend a growing proportion of their time in torpor, often changing roost site to facilitate this. Cool roosts are used, so that when they relax their body temperature to ambient levels, greater savings are made than in warmer roosts: a bat cannot lower its temperature below that of its environment. Bats preparing for hibernation can lay down 20–30 per cent of their weight in fat deposits in just a week or two. They do this not by eating more, but by going direct to cool roosts after foraging and dropping into torpor.

Hibernation

Successful hibernation may depend as much upon choosing the right roost as it does upon being physiologically prepared. Each species has its own particular requirements, but all species, to a greater or lesser degree, need a site that is cool, relatively humid, has a stable microclimate and is free from disturbance. Most British bats hibernate in caves, or artificial structures that fulfil similar requirements, such as mines, tunnels and cellars (Fig. 3.14).

Horseshoe bats, *Myotis* species, long-eared bats and the barbastelle all make extensive use of underground sites. Long-eared bats and barbastelles use trees and more exposed sites too, and when found in caves they tend to be near exposed and cool entrances. In fact, many of the cave-roosting species have different microclimate requirements and roosting habits. The temperature in a cave is buffered against the extreme fluctuations that may occur outside, but is to some degree related to the outside temperature. At the cave entrance, temperature may differ little from the outside, ambient temperature. Although the average temperature even quite deep in the cave may be correlated with ambient, the variation decreases with distance from the entrance. In the very deepest systems, annual variation may be no more than a degree or two or even absent. Temperature profiles will be more complex in caves with multiple entrances, and depend upon the relative heights of different parts of the cave and the entrances, and the prevailing winds outside. Even at a specific location, temperature will be lowest on the cave floor, and highest in blind crevices and depressions in the ceiling, where warm air will be trapped. The heat generated by roosting bats can further elevate the temperature in small depressions. The preferred hibernation temperatures of most species lie between 2 and 10°C. Larger species generally show a preference for lower temperatures. It is difficult and misleading to give precise figures, since preferred temperature often varies systematically through the winter and more erratically on a short timescale. For example, greater horseshoe bats tend to choose cooler sites in spring relative to winter, but during a given season, they will choose warm sites during a warm spell. This variation is linked to arousal frequency. They may choose warmer sites during warm spells so that arousal is triggered when it is warm enough for insects to be flying: greater horseshoe bats are known to feed during the winter. At a given temperature, arousal frequency increases as spring approaches, and the bats may move to cooler sites in the spring to reduce the number of unwanted arousals. Small species tend to arouse less frequently than large ones at a given temperature. At 10°C half a large group of greater horseshoe bats (25 grams) may arouse and change position in a single day. In a study of the Eastern pipistrelle, *Pipistrellus subflavus*, (6 grams) in a North American cave it took 44 days for half the bats to arouse and change their roosting position.

Some species are solitary roosters, others habitually hibernate in tightly packed clusters (Fig. 3.15). Most species fall between these extremes and the degree of clustering is related to temperature and physiological demands.

Fig. 3.14 Typical underground hibernation sites. From top left, clockwise: a sheltered limestone cave, an exposed limestone cave (with harp trap ready to place over entrance), a partially buried ice house and an adit to an extensive lead mine. The drawing at the bottom shows the many places bats roost.

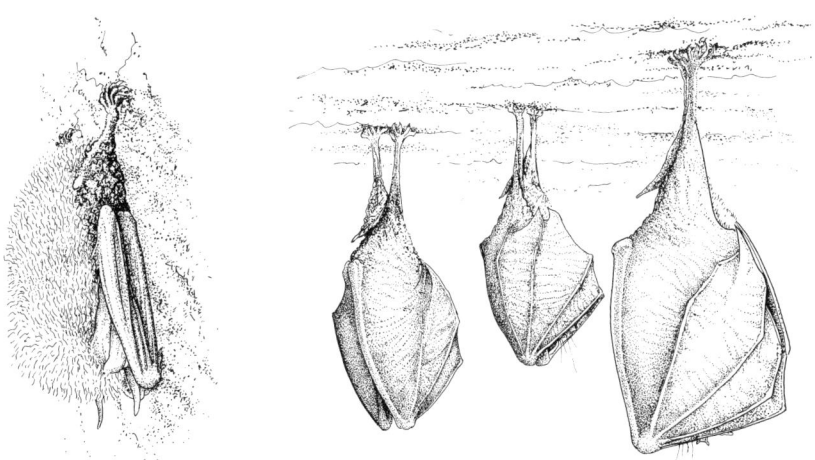

Fig. 3.15 Hibernating bats: a solitary long-eared bat (with ears tucked under wings) and a loose cluster of horseshoe bats.

Solitary bats have long bouts of torpor, arousing much less frequently than clustering bats of the same size. A large cluster of bats will maintain a higher body temperature than a solitary bat at less energetic cost, since heat retention is better. Clustering bats tend to occupy cooler sites in caves and form smaller clusters in warmer caves or in warmer weather.

Temperature selection and clustering behaviour therefore appear to be used to minimise energy expenditure and to regulate arousal. Patterns of arousal and the reasons why bats arouse are subjects of considerable debate. It seems that arousal is inevitable and probably even essential. Bats frequently arouse for no obvious reason: they may do no more than shuffle their feet and go back into torpor. As mentioned earlier, this may be to make adjustments to the many physiological processes that keep the body in order, including the need to mobilise the immune system to fight off infection. Body weight measurements show that greater horseshoe bats sometimes leave the roost and feed during hibernation. It has also been suggested that pipistrelles emerge from hibernation to feed, since emergence can be correlated with warm, insect-rich nights. Other studies, for example, on the brown long-eared bat, suggest emergence to drink may be more important. There is good evidence to suggest that clustering may be important in reducing water loss through the skin. Bats also arouse to mate: mating is known to occur throughout the winter in many species.

Arousal patterns may give some clues as to why a particular species comes out of torpor. The greater horseshoe bat is the only British species to be studied in detail in the wild. By using temperature sensitive radio transmitters and automatic logging of body temperature and echolocation calls, the bats could be monitored with minimal disturbance. Males and females (adult and immature) were shown to arouse at comparable frequencies and for a similar time. If the main function of arousal was mating, adult males should have aroused more frequently and for longer. The bats often remained in the cave during arousal, suggesting that feeding was not the primary reason for arousal.

Fig. 3.16 Arousal patterns of a hibernating greater horseshoe bat as spring approaches (from Park et. al, 2000).

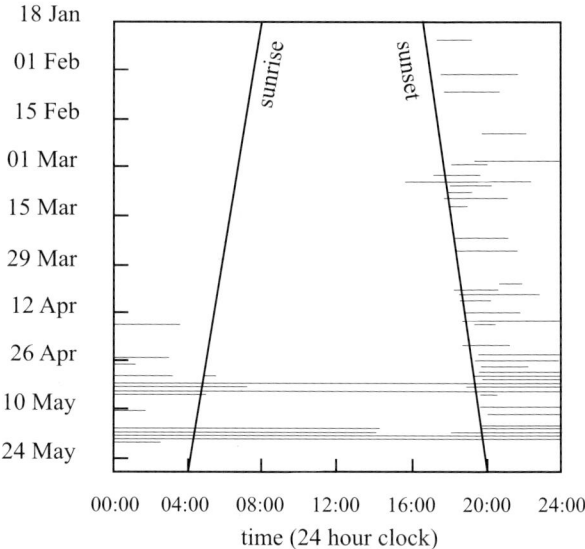

However, foraging did appear to be an important factor in determining the timing of arousal. Throughout the winter arousal occurred primarily around dusk, as expected if the bats are to have the opportunity to forage (Fig. 3.16). Significantly, the length of their active period after arousal increased only as temperature outside the cave increased above 10°C, when it is known that insects begin to fly in significant numbers. If the bats do feed, then they will need to remain active after they have returned to the cave to digest their meal. The bats appeared to select roosting positions where cave temperature, although buffered to some extent, fluctuated with the temperature outside the cave. By also synchronising their arousal with dusk, they are able to take advantage of warm nights to forage, even if that is not always the primary reason for arousal.

Very little is known about other British bats, but *Myotis* species in Europe and North America have been studied. In these, arousal in the wild is not correlated with sunset. Hibernating bats arouse at all times of the night and day, suggesting that arousal is not related to feeding. This makes some sense: in a cold, continental climate, opportunities to feed in winter are very few relative to those in the mild, maritime climate of Britain.

In continental Europe major hibernation sites attract large numbers of bats of many species. For example, Nietoperek in Poland (a large complex of underground, concrete tunnels) has over 20,000 bats of 11 species. Here in Britain, even the most important known hibernacula are much smaller. English Nature holds a database of almost 1,000 hibernation sites and some of the data are summarised in Figure 3.17.

Of the 991 sites, 666 of them have never had more than ten bats seen in them and only 42 sites have recorded more than 100 bats. Only four sites have recorded more than 400 bats and the maximum count for any site is just short of 700. Only one or two species have been recorded in 774 of the sites and only 63 sites have had more than five species in residence. Add together the maxi-

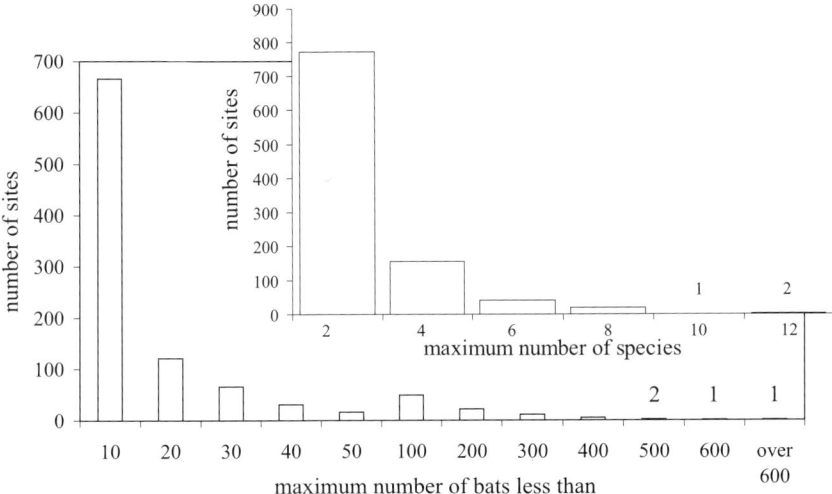

Fig. 3.17 Summary of data held by English Nature of known hibernacula: numbers of bats and species.

mum number of bats recorded in each site and the total is less than 21,100 of the estimated 500,000 bats that are primarily cave hibernators. The majority of the sites are man-made: only 21 of the top 100 sites are caves. Is this because the bats prefer man-made sites or because they come to our attention more readily in such sites and are then more easily located and counted? Horseshoe bats habitually hang free from the roof of a cave or mine, making it relatively easy to find them and monitor hibernation populations. Other cave-dwelling species show a variable but marked tendency to hibernate in cave cracks and crevices. If we isolate the rare but conspicuous horseshoe bats in our statistics what do we see? Figure 3.18 summarises data for the 93 largest sites, those with over 50 bats. Of these, 35 are occupied exclusively or predominantly by greater horseshoe bats, 22 by lesser horseshoe bats and ten have significant numbers of both species. The remaining 26 sites are dominated by *Myotis* species, only ten of which have counts exceeding 100 bats. Even more surprisingly, none of the top sites for Brandt's, Daubenton's, Natterer's or whiskered bats are caves. Known sites with significant populations of *Myotis* bats are all man-made. In contrast, 17 of the 67 most important horseshoe hibernacula are caves (Fig. 3.18).

To summarise, known hibernacula are predominantly man-made structures, such as mines, tunnels and cold buildings. They are largely occupied by rare but conspicuous horseshoe bats. Only a handful of sites are known to be home to *Myotis* species and brown long-eared bats, by far the most numerous cave-hibernators in Britain. So, the question is, where are the *Myotis* bats hibernating? I do not believe it is survey bias since there have been many organised cave surveys. I think we are simply failing to see most species because they prefer to hibernate in cracks and solution cavities. The greatest underestimates will be in natural caves that typically have many more cracks and crevices than mines. How many important sites remain undiscovered? We certainly cannot account

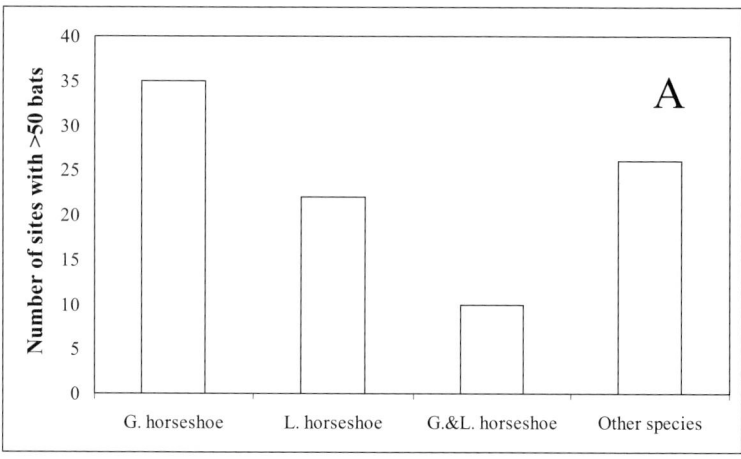

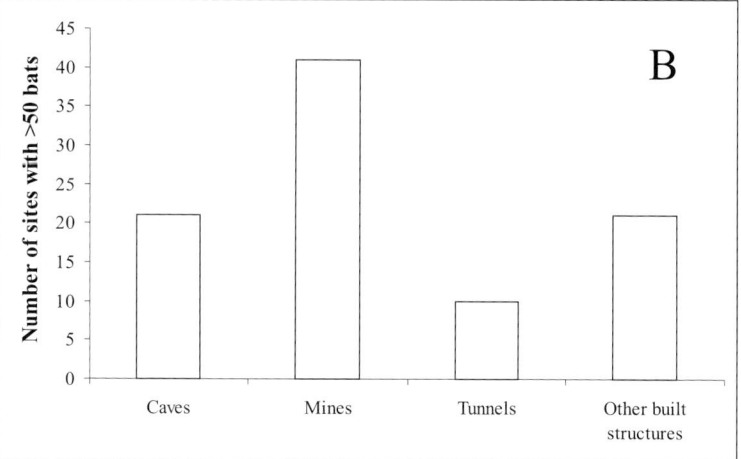

Fig. 3.18 More hibernacula statistics: dominant species and site characteristics.

for more than a tiny proportion of the country's cave-hibernators. Recent studies of autumn swarming suggest that hundreds or even thousands of bats visit some caves before hibernation. What proportion stay on to hibernate is not yet known, but I suspect it is very large. Visit these same caves in the spring and you will probably see significant numbers of bats emerge, yet winter searches typically fail to find the bats known to be there.

Noctules and Leisler's bats prefer tree holes, and serotines and pipistrelles appear to prefer buildings, although winter roosts are typically in cooler parts of buildings than those used in summer. Very little is known about hibernation in these bats, but their requirements do appear to be very different from those of cave hibernators. In the first instance, daily fluctuations in temperature and humidity must be very much greater than in caves. Energetically and physiologically these sites ought to be much more challenging to bats, but they have presumably evolved mechanisms to cope with it. Given the huge number of pipistrelles in Britain relative to all other species, it is remarkable that only a tiny fraction of this population can be accounted for in winter. A little more is known about noctules. Clusters of noctules in tree holes can tolerate temperatures as low as −16°C, when their surface temperature may be as low as −9°C. Between −9°C and + 10°C they arouse every 4–8 days, but at temperature above or below this range they may arouse every day. These results come from a captive colony in Europe, so we cannot be sure that wild noctules would not have left the tree hole to find a warmer site.

Hibernation and migration

In all temperate parts of the world, the onset of winter can make life difficult or impossible for many animals. As day length decreases and the sun's elevation in the sky declines, average temperature and the energy available for photosynthesis fall. Biological productivity falls at a time when homeothermic animals need more energy to stay warm. Birds and mammals have to make a choice: they can hibernate, migrate, change their diet, or store food. Bats usually hibernate, but it comes as a surprise to many people to discover that some, like birds, migrate. In North America, Mexican free-tailed bats migrate south from the United States to Mexico and Central America and remain active throughout the year. However, the majority of the species studied migrate not to remain active, but to find suitable hibernation sites. Tree-hibernating species, such as noctules and Nathusius' pipistrelles, migrate over much of their range and in continental Europe often fly south-west to more maritime climates, to avoid the more extreme inland continental conditions. Migrations of 1,000–1,700 kilometres have been recorded. Many other species migrate over shorter distances in more random directions to hibernate, and their movements probably reflect the local availability of caves rather than climatic gradients. Horseshoe bats are known to undertake short range winter migrations or dispersions in Britain, but virtually nothing is known of other species.

Knowledge of the movement of bats between summer and winter roosts has come from ringing studies. From about 1930–60 huge numbers of bats were ringed in their hibernation sites in Europe. This work also revealed that bats usually return to the same hibernation sites year after year. In part, this fidelity is due to the scarcity of sites with suitable microclimates. In Britain, natural caves are common in only a few areas, such as the Yorkshire Dales, South Wales and the Mendips. Even in these areas, only a small proportion of the caves are known to hold significant bat populations. In areas where natural caves are rare, bats use mines, tunnels and other artificial features, but what did they use before these structures existed? By taking advantage of these sites have they become less migratory?

Biological clocks

Biological clocks are internal controls that set the daily, seasonal or annual rhythms of various physiological processes. Although biological clocks are important in virtually all parts of a bat's life, they are at their most evident in studies of torpor and hibernation. What triggers hibernation? What controls arousal timing in cave-hibernating bats, in the absence of temperature and light cues? Biological clocks are built-in to bats, but they do not remain accurate unless they are continually recalibrated by environmental cues. A British bat arouses at dusk each summer night to take advantage of the peak in insect abundance. Its internal circadian (24-hour) clock therefore needs to be quite accurate. It stays accurate because it is reset by the daily light–dark cycle. Deprive a bat of this cue by keeping it in the dark and the clock begins to drift. *Eptesicus fuscus*, the North American big brown bat, has been kept in the dark at 2°C as part of a hibernation study. Under these conditions, the 'daily' clock period drifted to anywhere between 13 and 21 hours, although most bats continued to arouse within three hours of dusk. Just how well bats are able to keep time without external cues is still debated and is probably species-specific. However, it is clear that bats do use a circadian clock, regulated by the day–night cycle, to make sure they get up at the right time.

There is also some evidence for a circannual (12-month) clock. Pallid bats, *Antrozous pallidus*, from North America have been kept under constant photoperiod (day and night duration), temperature and food supply for over two years. Despite this, each autumn they put on weight in readiness for hibernation. The circannual clock is probably regulated by photoperiod and temperature. During the autumn the nights get longer and colder and trigger pre-hibernation behaviour. It makes sense to have a clock that tells bats when winter is approaching, so they are not completely misled by any temporary, unseasonable changes in weather or food supply. However, bats have an ability to override tight control depending upon short-term variations in weather, food supply and other items on their agenda, such as mating.

Life history cycles and reproduction

Temperate bats have a diversity of social structures and mating strategies that is unusual in mammals of their size. Their life history strategy is one of long life expectancy and low reproductive rate, when that of most other small mammals is one of brief fecundity (p. 28) Mortality and longevity of wild bats have been studied in several species, and there are anecdotal records of age from many more. The recorded maximum age of all British species ranges from 7–30 years, with an average maximum of about 20 years. The annual survival rate for brown long-eared bats is about 75 per cent, with many bats surviving for 15 years, although females tend to live longer than males. In greater horseshoe bats annual survival increases rapidly from around 50 per cent in the first year to 80–90 per cent by the age of six, with an average of over 70 per cent. Limited data suggest that most species have an average life expectancy of over four years and that mortality is high in the first year, but diminishes rapidly thereafter. It is not known why bats live so long, but flight may make it necessary. Young bats are born large and must be virtually adult size before they can fly. They must then be weaned and achieve independence in a short summer season. There is a limit to the number of large offspring a bat can produce at one time: usually one, occasionally two and very rarely more. If bats cannot produce many offspring, they must have low juvenile mortality and long reproductive lives to produce sufficient young to replace natural losses. This long life, in combination with communal roosting, mobility and other aspects of their biology, provides the flexibility for a diversity of social systems.

The annual cycle

An overview

Figure 3.19 shows the stage in the life cycle of a typical British bat at any particular time of the year. The figure can only be an approximation, since so

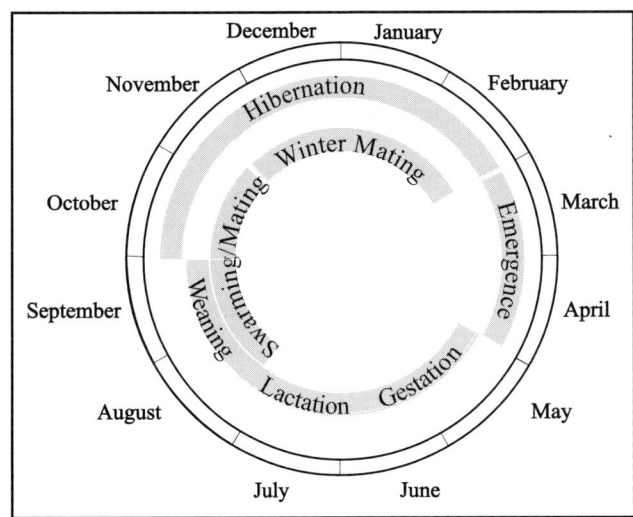

Fig. 3.19 The annual cycle.

much depends upon the weather and the physiological status of the bats. However, it does give an overview of a year in the life of an average bat. The year begins with the end of hibernation, which may be any time through March, April and often into May. Even in the same year different individuals, within and between species, may emerge from hibernation at different times. Nursery colonies are formed around the beginning of May and usually break up some time in August or September. Gestation in May and June typically leads to birth in late June and July with weaning from late July through August. As nursery colonies break up in August and September, mating occurs during the swarming season and in the early stages of hibernation. As winter progresses, mating becomes less frequent.

Emergence from hibernation and summer roost formation

Bats begin to emerge from their hibernation sites in significant numbers from March onwards, depending upon the weather and the food supply. As the spring progresses they move from their hibernation sites to summer roosts in trees and buildings. Some of these roosts will be transitional. They may have microclimates more suitable for intermittent torpor during the fickle spring weather. They may also serve as gathering places for communal roosters, before these move on to the summer roosts proper. Males of most species studied to date will spend the summer roosting alone, or in small groups with other males. Females tend to form larger nursery colonies. However, there is considerable variation. The number of males present in brown long-eared bat nursery roosts increases through the summer, and they may make up 30 per cent of the roost by September. Males in significant numbers (20 per cent or more) are found in nursery roosts of greater and lesser horseshoe bats, long-eared bats, Natterer's bats and Daubenton's bats. Nursery roosts of 45 and 55 kHz pipistrelles are often made up entirely of females and males rarely make up more than 1–2 per cent of the bats in the roost. Males are also rarely found in the nursery roosts of other British species, although data are sometimes only available from sites in continental Europe. Nursery roosts are typically made up of ten to 100 bats, sometimes up to about 200, and, in a few species, notably pipistrelles, in excess of 1,000.

Summer roosts, and particularly nursery roosts, are typically in relatively warm and dry locations in trees and buildings (Fig. 3.20). Horseshoe bats, pipistrelles and serotines are rarely found roosting outside buildings in summer. Pipistrelles will use bat boxes early in the season, solitary bats are found in tree holes and bark crevices, and males will use bat boxes as mating roosts. Some species, such as Bechstein's bat and noctule, are predominantly tree roosters, but occasionally use built structures. Other species are very adaptable. The list of roosting sites used by bats is almost endless and virtually any secure crevice or cavity may be exploited. Horseshoe bats must be able to fly into the roost and direct to their roosting site. An acrobatic turn in flight deposits them on the spot. They often enter buildings through windows, doorways, ventilation shafts and other reasonably large openings. Other species land on the outside of the roost and crawl in, often through very small openings. Horseshoe bats, and to some extent long-eared bats, roost in the open, but most species prefer smaller crevices, even within a large roof space.

Despite the great variation seen in roost sites, the basic requirements are similar for all species. The roost must give security from predation and provide a microclimate suitable for minimising thermoregulatory costs and for

Fig. 3.20 A cluster of alert Daubenton's bats in a summer roost.

rearing young. Some species, such as horseshoe bats, appear to have very exacting requirements: a contributory factor to their recent decline and a challenge to their conservation. A good roost site should also be close to good foraging sites (p. 89).

From conception to weaning

Having established themselves in the nursery roost, the females prepare for birth. Mating may have occurred at any time during the autumn and winter, but pregnancy is delayed. By mating in the previous autumn and winter, bats ensure that pregnancy can begin as soon as conditions are favourable. Bats that are born early, grow quickly and are weaned early have a much greater chance of survival. In all British bats, the egg is retained in the ovary after copulation and the sperm are retained in the oviduct until the spring. There is some evidence to suggest that the sperm are actually nourished in the oviduct to maintain their viability. Sperm production in the males stops between September and November, so the male too must be capable of storing sperm if mating is to occur through the winter. In the spring, when the females are physiologically prepared and the food supply is adequate, the egg is released and fertilised, and the embryo implants in the uterus. Given an adequate food supply, gestation in a fully homeothermic pipistrelle is around 40–50 days in average years. It is longer in the larger long-eared bats, at 60–70 days, and 70 days in the noctule. A poor supply of food in combination with low environmental temperatures can lengthen gestation by 20 per cent or more, since there is too little food to nourish the growing foetus and to maintain homeothermy for rapid foetal growth. Females are usually homeothermic throughout all but early pregnancy unless the food supply is poor. Female greater horseshoe bats have been known to use heterothermy just prior to birth. It has been suggested that this may be an attempt by some females to synchronise births. In very bad years, females may find it necessary to abort their foetus to secure their own survival.

British species usually produce a single offspring and twins are rare. In continental Europe, some of the same species produce twins more often. Pups are born between late June and early August. Births are generally highly synchronised: 70–100 per cent of the females will give birth, and all pups will usually be born over a 2–3 week period. Most bats give birth in a head-up position, perhaps cradling the baby with the wings and tail membrane as it is born. At birth, pups are as much as 20–30 per cent of adult mass (compared to 5–10 per cent in other mammals of similar size). This high birth weight is due to the unusually long gestation period in bats relative to other mammals of a similar size, since foetal growth rates are very slow. Total litter mass is comparable to that in other small mammals, which usually produce more young. Why do bats produce just one pup? The most plausible hypothesis is related to calcium requirements. Young bats must reach near adult size and have well-calcified, strong bones before they can meet the aerodynamic and mechanical requirements of flight. Bats have a low calcium diet and there is some evidence that females must mobilise some of their own body calcium to meet the needs of their pups. Producing more than one large pup is rarely an option. When bats do produce twins, the combined weight is usually high and the degree of development similar to that of single pups, suggesting twins are only born under particularly favourable circumstances.

Pups are blind and naked at birth, but the eyes open within the first few days and fur grows rapidly (Fig. 3.21). They have milk teeth at birth and quickly learn to grip the mother and suckle. The incisors are curved and grip the nipple firmly. Pups often have disproportionately large feet for climbing and gripping. Growth is rapid, with pups reaching 70 per cent of adult mass and over 90 per cent of adult skeletal dimensions in as little as 14–22 days. Only at this stage do the wings reach adult size and proportions, allowing the pups to make their first flights. Prior to this the wings are small, with a disproportionately large arm-wing, and the wing loading is too high for flight. Pups begin to feed themselves at this stage and will be weaned about 45–65 days after birth.

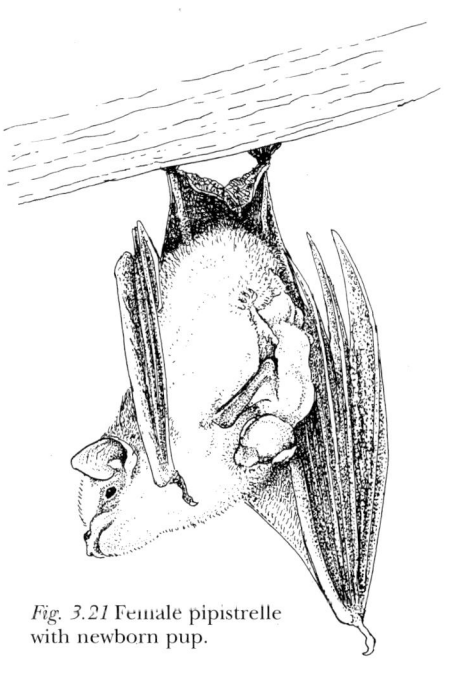

Fig. 3.21 Female pipistrelle with newborn pup.

Growth rate and early mortality are determined by the quality of the food supply, the microclimate of the roost and the age and experience of the mother. The pups most likely to survive are those born to experienced females, in roosts where thermoregulatory costs are minimised and during periods of high insect abundance. Reproductive success of female greater horseshoe bats increases up to the age of eight. No information is available for other species, but it is reasonable to assume that as females of all species gain in experience, reproductive success is likely to improve. Very young bats do not thermoregulate, so low environmental temperature means low growth rates due to low body temperatures and slow metabolism. Once pups do begin to thermoregulate, resources that could be put into growth may be used to keep warm and this too will reduce growth rate in cool roosts. Most of our native bats rear their young in roosts that receive a significant amount of heat from the sun and are structured so as to retain this heat. Heat generated by the bats themselves may also be important. There are also many examples of bats exploiting artificial heat, for example, from central heating systems and the ubiquitous solid fuel cookers and heaters of fashionable country cottages.

Females appear to make greater use of torpor during lactation than during pregnancy. This has led to some debate about the relative costs to the female of pregnancy and lactation. I suspect that variation in environmental conditions, food availability and possibly species-related foraging and roosting patterns make it difficult to generalise. Milk production need not be reduced by the use of torpor. Suckling induces rapid milk production, topping up supplies during periods of homeothermy, giving lactating females more flexibility in their energy budgeting than pregnant females.

Feeding patterns

Foraging behaviour of females is also variable. A pattern observed in pipistrelles can be used as a starting point. In early pregnancy most of the bats leave the roost at dusk, but within an hour begin to return at a steady rate until all are back in the roost by dawn. As pregnancy progresses, most bats are out of the roost for most of the night and return towards dawn. During lactation and weaning many bats return to the roost during the night, presumably to suckle their young. In the hours leading up to dawn, most bats are again out of the roost. Post-weaning, bats forage primarily in the hours just after sunset, returning to the roost at a steady rate through the rest of the night. This pattern has been confirmed for individual, radio-tracked Daubenton's bats. Pregnant females rarely return to the roost, but lactating females return once and sometimes twice each night, for 20–60 minutes each time. Post-weaning foraging patterns become less predictable, but are frequently confined to the first few hours after dark. Bats may forage for as little as one hour on some nights. Lactating brown long-eared bats also return to the roost several times during the night to suckle.

Mother–pup interactions

Pups are usually found firmly attached to their mothers, but when they go out to forage, females leave their young in the roost. Carrying pups increases the cost of flight and reduces manoeuvrability, so pups are usually only carried when the female is changing roost. The pups typically form clusters when the females are absent. Females suckle only their own pups even in the largest roosts with the largest crèches. This requires mutual recognition between mother and pup, which must be established within hours of birth. Vocalisation plays a key role in the bonding and subsequent recognition process. Within minutes of birth, pups emit what are known as isolation calls: short calls, with ultrasonic components, which often resemble parts of the adult echolocation call. In pipistrelles, and probably most British species, as the pups grow, these isolation calls become more complex and resemble echolocation calls more closely. The females also emit specific calls when establishing contact with their pups and respond specifically to the isolation calls of their own young. Work on North American species such as the evening bat, *Nycticeus humeralis*, and Mexican free-tailed bat, *Tadarida brasiliensis*, shows that these calls are sufficiently varied between individuals, and reproducible within individuals, for mother and pup to identify them. Smell is also an important factor in mother-pup recognition in pipistrelles and probably most other species.

Weaning and weaned greater horseshoe bat pups forage independently of their mothers, but over the first year of life, survival of the pup depends to some extent on the survival of its mother. This suggests that the pup may learn some of the skills of life from its mother. Recent work shows that mother and offspring share both roosts and foraging areas over several years. Since individuals forage alone, the benefit to the offspring may be the inheritance of high quality roosts and foraging areas, rather than close co-operation with its mother in locating food and feeding. Males have no role in pup rearing in any temperate species studied.

Sexual maturity

Most British bats, males and females, become sexually mature when about one year old. A small proportion reach maturity in their first autumn at three months old. In some species, for example, long-eared bats, females may not reach maturity until they are two or three years old. Some of our larger species may take even longer: male greater horseshoe bats usually become mature at two years and females at three, but there is considerable variation.

Mating patterns

British bats mate from late summer and mating may continue through the winter. Bats are wonderfully diverse in their social systems and mating behaviour, and even the few British species studied show a surprising diversity. Underlying this diversity is a single goal: maximising reproductive success. Since males produce many sperm and take no part in rearing the young, their strategy is largely one of mating with as many females as possible. Choosing females with a high 'fitness', in other words females most likely to produce healthy offspring, may also be a factor, as may the exclusion of other males. Females must select healthy and dominant males: those most likely to pass on genes that give her offspring the best chance of surviving and reproducing. Since females produce only one or two young, and have to invest considerable time and energy in rearing their offspring, selecting the right male is important.

For most of the year, males are concerned only with staying alive. Females, on the other hand, have to rear their pups during our short and fickle summer when food is most abundant. The complex demands imposed on the females may to some extent be met by complex social behaviour. For example, by roosting together, females may be less vulnerable to predators, have lower thermoregulatory costs for themselves and their pups and be able to share information about good foraging sites. Group foraging may also bring benefits. The last two will be discussed further in Chapter 4 (p. 93). The mating strategies evolved by the males must fit into the social framework of the females if they are to be successful. With these considerations in mind the various patterns seen in British bats may be described and perhaps explained, and the most likely strategies to be used by those species that have yet to be studied may be predicted.

Few bats are monogamous, and all British bats are polygynous (males mate with several females). This is not surprising: since males do not help to rear the young there is no evolutionary advantage in remaining faithful to a single female: more mates means more offspring. Polygyny comes in many forms. At one extreme, males may defend females or groups of females from other males. Alternatively they may defend territories and roosts. Territories or roosts that offer good feeding or 'housing' to females may encourage them to stay and to mate, and males may chase other males from the site to keep the females to themselves. Temporarily held mating sites or lekking sites may be held by males for a brief period in the autumn, with the aim of attracting females. Finally, mating may be promiscuous. Males may act alone in all these situations, but may also form alliances with other males.

Defending a resource, such as a territory, or females themselves, makes good sense. By monopolising females in some way the males ensure that they alone father any offspring. However, it is not always feasible. Females may be scattered over too great a home range, or live in groups that constantly change composition. Males would find such groups difficult or impossible to defend. Even if it is possible, the advantages gained may not justify the time and energy spent. An alternative is to defend a roost site. A safe roost site close to good foraging, particularly if it is a rare resource, will attract females and is therefore worth defending. However, even this may not be justified if the females are particularly nomadic and make use of several roosts. The males of lekking species defend small, temporary display sites, from which they attract passing females, with no attempt at subsequent defence. In some species mating may involve no evident signs of display and selection. In all but the latter, the male must have something to offer the female: himself or some desirable resource such as a good feeding site or roost site. The female has the choice of whether to accept or reject what the male has to offer. So, that is the theory, but can the behaviour of our native species be fitted into this framework? Too little is known about most of our bats even to speculate, but recent research is beginning to provide some clues.

Without doubt, the best studied species is the greater horseshoe bat. Females form nursery colonies of varying sizes, sometimes large, usually in buildings that are used year after year. Males are usually solitary and often roost underground, even in the summer. Mating usually occurs in the autumn, but can continue right through to the spring. It is the females who visit the males, the latter holding individual territories in caves, mines, cellars or other underground sites. Males have been shown to hold the same territory for up to 16 years, and as many as eight females have been seen to visit a single male in one season. After mating, a vaginal plug forms in the female that may prevent further mating. Breeding therefore appears to be polygynous with only a proportion of males being successful, perhaps over many years. Recent genetic studies on a well-known colony support this view, but mating success was not as skewed as behavioural studies suggested. This may be because females can reject their vaginal plugs, are known to disperse over a wide area and can continue to mate throughout the winter and spring. It is not yet known whether females choose specific males or whichever male happens to be holding a favoured territory.

From studies using bat boxes in Britain and continental Europe, it is reasonable to suggest that both 45 and 55 kHz pipistrelles operate a system of resource-defence polygyny. Male pipistrelles establish themselves in solitary roosts during the summer. Late in the summer, the nursery colonies fragment as the pups of the year are weaned, and the females and their offspring disperse to many smaller roosts. At this time, males appear to advertise their presence to females with a song flight. They fly backwards and forwards in the vicinity of their roost, emitting a distinctive multisyllabic call every few seconds (Fig. 3.22). The calls are broadband, reaching down to 14 kHz, and are audible to many humans. Although direct evidence to support the idea is lacking as yet, it is presumed that it is this display that attracts the one to ten females known to share the roost with the male. During the autumn mating season, males do not share the same roost and do not appear even to tolerate other males of the same species in nearby roosts.

Nathusius' pipistrelle also uses a song flight, or calls from a stationary position outside the roost. Male noctules call from outside the roost too (Fig. 3.23), using a low frequency call clearly audible to humans. Mating calls may be low

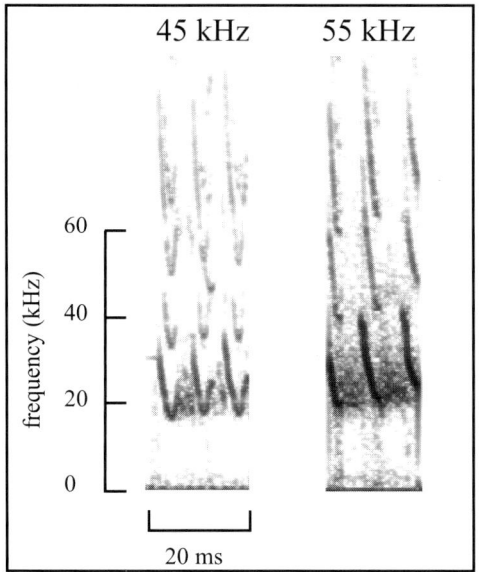

Fig. 3.22 Pipistrelle mating calls. The differences between the calls of 45 and 55 kHz pipistrelles are not as obvious as the selected calls suggest.

frequency to limit attenuation, so that they travel further through the air and attract mates from a large area. Many other species are known to emit social calls, but it is not known whether these are used specifically in mating.

Nursery colonies of brown long-eared bats often contain adult males in late summer and mating occurs at this time. Female long-eared bats form stable groups that may be relatively easy to defend against other males, at least at this time of the year. The number of males present increases with the number of females: it is possible that males are defending groups of females and as group size grows, more males take on the task of female defence. However, as in many other species, mating can occur from late summer right through to the spring. Female defence by males is unlikely to occur during hibernation. This suggests that at least some of the mating is more promiscuous. There is no reason why different mating strategies may not be used at different times of the year.

Fig. 3.23 Male noctule calling from his tree roost to attract mates.

Nursery roosts of both Natterer's bats and Daubenton's bats contain males and mating may occur in these roosts late in the summer. Since they are also common at swarming sites (see below) and both are known to mate during hibernation, it is possible that mating occurs over a long period, and more than one mating strategy may be operating. Male Daubenton's bats, in addition to roosting in nursery colonies, form large all-male roosts of more than 60 bats. In upland areas in Britain and Europe, these males forage along the upper stretches of rivers, and females may be absent. Small numbers of males may roost with the downstream nursery colonies, but the extent to which mating occurs at these sites is unknown. Is mating largely the prerogative of the minority of males that roost and feed downstream in summer, or is mating at swarming and hibernation sites more important, when both male populations may be present? These questions cannot yet be answered, but research is underway that may provide solutions. These complex issues are almost certainly related to other aspects of bat behaviour. The next chapter broadens the discussion and attempts to integrate many of these aspects.

Swarming

A spectacular but enigmatic seasonal event known as swarming almost certainly has a mating element. Although swarming has been known about for many years, particularly from studies in North America, until very recently it has been largely neglected in Britain. However, once observed swarming is hard to forget and efforts are now being made to document and understand it. Why it happens is a fascinating question in itself, but it may also be important to bat conservation. At this stage I can speak only about my own experiences, with a little feedback from colleagues: swarming studies are still in their infancy and there is still a great deal of guesswork involved.

Put simply, swarming involves the gathering of large numbers of bats, often of several species, in and around the entrances to caves and mines. Swarming can be observed from August to November, often with a relatively brief period of maximum activity, typically in September. The timing may vary from year to year and is also determined by locality and species. Swarming sites are usually also hibernation sites, although many more bats may swarm at a site than subsequently hibernate there, but this is not known for certain. Several adjacent sites have been monitored in a number of localities in the Yorkshire Dales and North York Moors. Although all are known to be hibernation sites, only a small subset appear to be major swarming sites. Even in August a small number of bats may roost underground and leave at dusk. However, activity does not peak until several hours after dark and may continue for several hours more. As the autumn progresses a dusk activity peak become increasingly evident, due to the resident cave bats, but it rarely challenges the enormous later peak composed of bats gathering from other localities. All the evidence suggests that bats visit key swarming sites from a very large surrounding area and that particular bats visit only briefly. Hundreds of Natterer's bats alone have been captured and ringed over the last four years at one site in North Yorkshire. It is too early to make an accurate estimate, but preliminary recapture data suggest a visiting population of several thousand Natterer's bats.

Swarming sites are dominated by *Myotis* species, particularly Natterer's and Daubenton's bats, with Natterer's bats generally being most common. However, this may in part reflect the timing of visits: Daubenton's bats appear to have a peak of activity earlier than Natterer's bats. Whiskered, Brandt's and brown long-eared bats are also present, but as yet make up a small minority. In the north of England these five are the only common species known to use caves for hibernation on a regular basis, so we would not expect to find others. On a given night from three to all five are likely to be caught. More cave-roosting species are found further south, and on a memorable visit to a mine in the Southwest I caught eight species in one evening, adding greater and lesser horseshoe bats and the rare Bechstein's bat to the above list. At other sites barbastelle bats have been recorded.

So, why do bats swarm? The bats repeatedly circle, often in large aggregations, in the cave or mine and around the entrance, frequently chasing or shadowing each other (Fig. 3.24). In North America swarming bats are evidently involved in mating within the cave. Although mating appears promiscuous, there is some evidence that mating success is biased towards a minority of males. If mating is also a major function of swarming in Britain, then there are some puzzling details. Males typically make up 80 per cent of the bats present on a given night. Over many nights recording, relatively few social calls have been heard, an unexpected observation if the activity has a strong social element. Nevertheless, mating remains the most likely major function for swarming and it may be an important mechanism in maintaining genetic diversity within bat populations, since the bats appear to travel far beyond their summer ranges. Interestingly, the only study that has looked at sex ratios of hibernating *Myotis* bats (Brandt's and whiskered) found that males outnumbered females three- to six-fold. Even if mating is the most important function, it does not preclude others. It may be the way young bats learn the locations of important underground sites. Individual bats can visit more than one cave

Fig. 3.24 Bats swarming at a cave entrance.

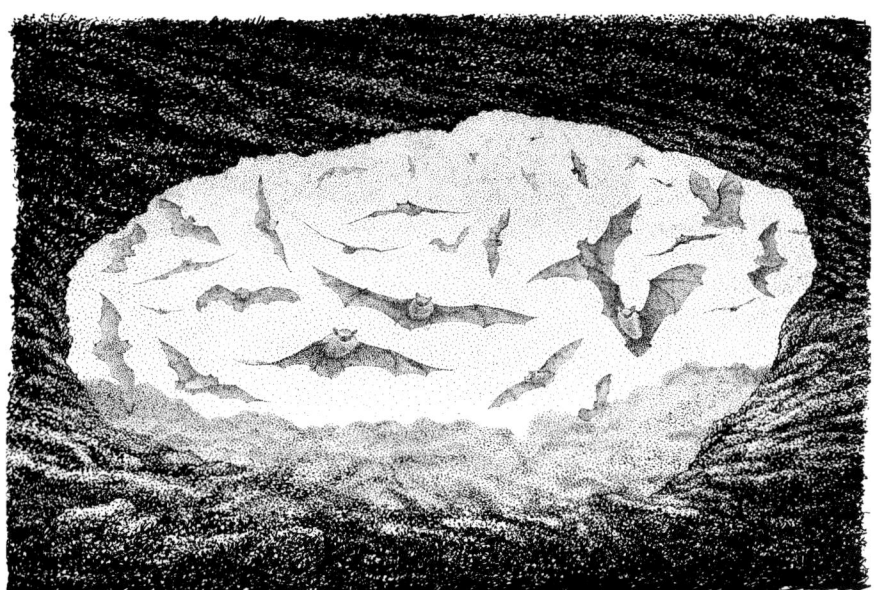

in a single evening. Whatever its function, swarming is clearly an important stage in the annual cycle of many bat species and research into this activity is clearly a priority. If major swarming events occur mainly at a relatively small number of sites, protection of these caves and mines may be an important conservation issue.

4

An Ecological Synthesis

The last chapter considered different aspects of bat biology, but all were discussed in relative isolation. Of course, in reality they are all interrelated. Foraging strategy, echolocation call structure, wing shape and flight patterns all evolve together to equip a bat for its particular way of life. What bats eat also determines to some extent their social lives and their patterns of reproduction. What they eat has been determined not only by what food is available, but also by what else is eating it. Another important determinant of ecology and behaviour is what is eating the bats. Unravelling these interactions is one of the greatest challenges in bat biology, conceptually and practically, because of their complexity and circularity. This chapter discusses some of the progress made. Perhaps the most intuitively obvious interrelations are those between flight, echolocation and feeding, and that is where the chapter begins.

Wing form and flight

We are all familiar with the fact that the wings of aeroplanes come in many shapes and sizes: gliders have very different wings to fast jets. This is due to the flight performance characteristics required and reflects their different aerodynamic properties. Very similar aerodynamic rules apply to bats (Fig. 4.1). A bat's wing can be described in several ways, but in basic terms the descriptions reflect the area of the wing in relation to the weight of the bat and the wing's shape. Two simple measurements can sum up much of the variation in the wings of different bats – wing loading and aspect ratio – and these can be shown to have a major influence on flight performance, that can in turn be related to the ecological niches occupied by the bat.

Wing loading and aspect ratio in relation to foraging style

Wing loading is simply the weight of a bat divided by the total area of its wings. Bats with a high wing loading are large and heavy in relation to their wing area, bats with small bodies and large wings have a low wing loading (compare the bats in Figure 4.1). Intuitively, it is clear that bats with a high wing loading will find it harder to stay airborne, since their wings will not generate sufficient lift. To compensate for this, bats with a high wing loading have to fly rapidly. In contrast, bats with a low wing loading can stay airborne at low flight speeds. However, wings with low areas not only generate low lift, they also produce low drag. Because less energy is expended in overcoming drag, bats with a high wing loading fly more efficiently. High efficiency in this context means less energy is expended in flying a given distance or for a given time. Since this energy must come from the flight muscles, powered by the food the bat eats, high efficiency is ecologically important. As always, there are trade-offs. A fast-flying bat is not very manoeuvrable: a high flight speed means a large turning circle. Fast-flying bats are not good at flying in dense woodland, for example. A high wing loading also makes take-off more difficult.

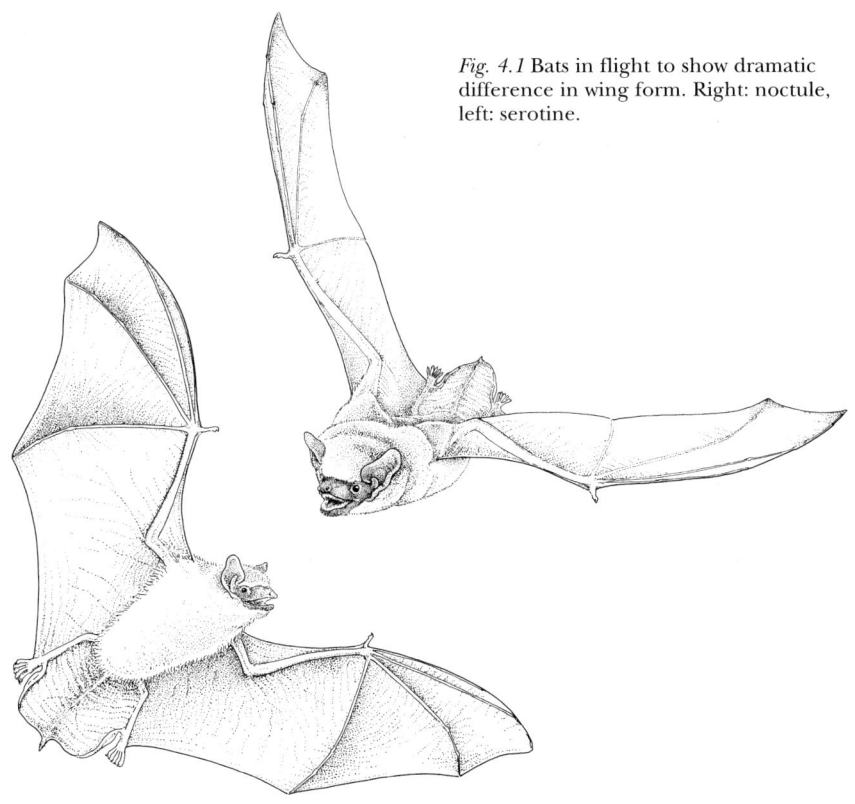

Fig. 4.1 Bats in flight to show dramatic difference in wing form. Right: noctule, left: serotine.

Aspect ratio is used to describe the shape of a bat's wing. It is the ratio of wingspan to average wing width. Since bats have such an irregular wing shape, aspect ratio can be defined more accurately as wingspan squared divided by wing area. A high aspect ratio wing is therefore long and thin, a low aspect ratio wing is short and wide (Fig. 4.1). The main benefit of a high aspect ratio wing over a low aspect ratio wing of the same area is increased aerodynamic efficiency: less energy is wasted. The major disadvantage of high aspect ratio wings is that their length makes them cumbersome in confined spaces and during take-off.

Even with just these two measurements, we can begin to categorise bats according to wing shape and make and test predictions about what sort of lifestyle or ecology the bats have. Figure 4.2 plots wing loading against aspect ratio for all the British bats. Bats with the lowest wing loading and aspect ratio are in the bottom left hand corner, those with the highest in the top right. A little caution must be exercised in interpreting this graph, since other factors influence flight performance. Furthermore, wing shape may be determined by factors other than flight performance. To this must also be added natural variation between populations. Wing loading data are available from several sources for some species and the values shown in the graph are means. This

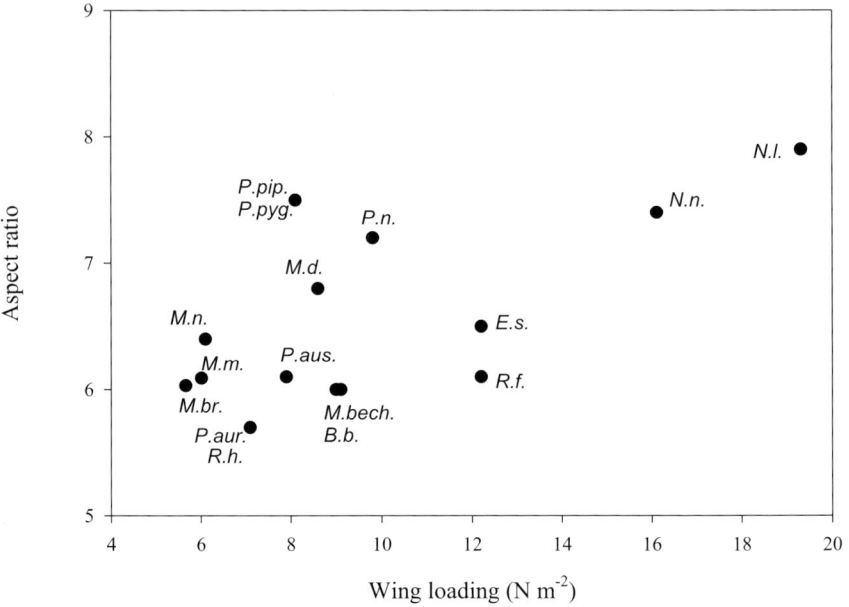

Fig. 4.2 Wing loading (in Newtons per square metre) plotted against aspect ratio for British bats.

variation does not alter the basic conclusions, but is a reminder that too much should not be read into small differences. Despite these reservations, significant conclusions may be drawn from this graph.

Long-eared bats have a low wing loading and low aspect ratio wings. Wings of this size and shape allow the bats to be slow and manoeuvrable fliers, and facilitate easy take-off. On the other hand, flight is not particularly efficient. These characteristics describe long-eared bats very well. They hunt by flying in dense vegetation, taking prey in the air, but also by gleaning from the trees and the ground. To do this they must be slow and manoeuvrable: flight speed has been measured at about 3 metres per second (ms^{-1}). Flight speed is very difficult to measure in the field, so the values given are only a guide. Since they frequently land, long-eared bats must be able to take off with the minimum of effort, both to save energy and to escape ground predators. Long-eared bats frequently capture heavy prey, such as large moths, which are too big to be eaten quickly. They habitually take large prey to a favourite feeding perch, where they remove the insect's wings before eating the nutritious parts. Their low wing loading gives them spare carrying capacity, enabling them to take off and fly with these heavy payloads. Long-eared bats rarely forage far from their roost. Since they do not fly long distances, they can afford to sacrifice high flight efficiency for these benefits. The drag on their enormous ears must also decrease flight efficiency, but these ears allow them to detect the low intensity rustlings of their fluttering or walking prey. Long-eared bats frequently hover. Aerodynamically, hoverers should have long wings to increase their efficiency: the more air swept down by the beating wings the better. Yet virtually all hovering bats have short wings: the most likely explanation is that long wings would be a physical hindrance in a confined space and most hovering bats hover near vegetation or the ground to feed. Care must be taken not to fall into the trap of thinking that aerodynamic considerations are always the most important.

Both Natterer's and lesser horseshoe bats are known to have foraging habitats similar to long-eared bats (flight speeds are 4.5 and 3.5 ms^{-1} respectively). Their position on the graph near to long-eared bats is therefore not surprising. Natterer's bats have been described as both hawkers and gleaners in a number of studies and their diet reflects this behaviour. Very recent work shows that they have an additional strategy. Unusually (so far as is known), they routinely catch insects very to close to vegetation *without landing* (Fig. 4.3). Prey are taken from distances of 5–10 centimetres from vegetation and often as little as 2 centimetres. This must require flight skills at least as demanding, and prob-

Fig. 4.3 Natterer's bat gleaning spider from web.

ably more demanding, than gleaning. The lesser horseshoe bat also has a versatile strategy, taking prey by hawking and gleaning.

Very little is known about the foraging behaviour of the closely related whiskered and Brandt's bats, at least from direct observation. Their position on the graph indicates a wing morphology suited to slow, manoeuvrable flight and the speed of whiskered bats has been measured at 4.3 ms^{-1}. This suggests they too are woodland foragers or at least that they forage close to vegetation. Brandt's bats feed largely on night-flying flies and these are most probably taken in flight, but a significant component of the diet is made up of diurnal flies and spiders, suggesting that some of their prey is gleaned. The diet of the whiskered bat is very similar. The few observations made of these bats feeding are consistent with a slow flying and gleaning foraging strategy.

Nearby on the graph, with only slightly higher wing loadings, are Bechstein's and barbastelle bats. Both species forage predominantly on moths, usually in or around dense woodland, and their strategies, although slightly different, require slow manoeuvrable flight (Bechstein's 4.9 ms^{-1}). Bechstein's bat feeds by aerial hawking and by gleaning: non-flying arthropods appear in the faeces of many individuals. As yet, few direct studies have been undertaken, but a study in Dorset found that they preferred closed-canopy areas of mature deciduous forest. Studies of barbastelle bats suggest an even greater dependence on moths, taken during slow flight, with as yet no direct evidence for gleaning. One study indicates foraging above the tree canopy, but observations were admittedly few. They are much less likely to be observed when foraging within or below the canopy, a behaviour observed in radio-tracked bats in Sussex. They appear to do this just after emergence when the evening is still light, often moving into the open as darkness falls, but still foraging largely in the vicinity of woodland. Studies in Somerset reported radio-tracked bats feeding in woodland, scrub and wood pasture, with bats frequently moving in the open: the barbastelle appears to be a very versatile feeder.

The three pipistrelle species and Daubenton's bats have a similar wing loading to the woodland species. Perhaps more importantly, they have significantly greater aspect ratios. The longer, narrower wings confer greater aerodynamic efficiency for a similar wing loading. Similar wing loading also suggests they will have similar flight speeds and should therefore be relatively manoeuvrable. The limited data available suggest that pipistrelles and Daubenton's forage at a flight speed of about 4–4.5 ms^{-1}, compared to 6–7 ms^{-1} for noctules and 3 ms^{-1} in the case of long-eared bats. Pipistrelles feed predominantly by catching small insects in flight, along tree lines, hedgerows and woodland clearings. Both the 45 and 55 kHz pipistrelles feed in the vicinity of water, but the 55 shows a greater preference for water. Daubenton's bats almost invariably feed low over open water, preferring tree-lined, still water. All these species benefit from increased flight efficiency, and will not be disadvantaged by their relatively longer wings.

Serotines and greater horseshoe bats have relatively low aspect ratio wings, but their wing loading is significantly higher than that of the species discussed so far (serotines have been recorded as foraging at 6.2 ms^{-1}). This may in part be due to their greater size, since wing loading increases with size for bats of similar proportions. This is because mass increases faster than area (they are proportional to length squared and cubed, respectively). However, this is clearly not the whole story, given the relative positions on the graph of some of the

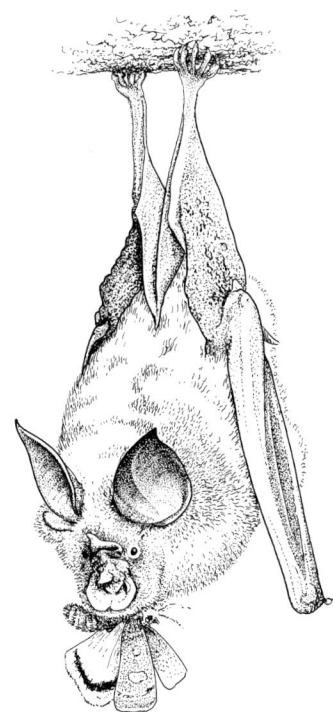

Fig. 4.4 Greater horseshoe bat flycatching.

other species. Both are typically open habitat foragers. The serotine is a primarily an aerial hawker, but it has an aspect ratio that is low for a bat of open spaces. Part of the answer may be that although the serotine eats beetles, moths and other insects caught in flight, it also feeds on some diurnal insects and has been observed gleaning. It frequently flies low over pastures and meadows, but will feed in a wide range of habitats. The greater horseshoe bat also uses a wide range of habitats, but has a preference for pastures and woodland. When feeding in pastures it flies relatively close to the ground and along hedge lines, catching insects by aerial hawking, but it will also glean and hunt from perches by flycatching (Fig. 4.4), taking short forays from its perch to intercept passing prey. This is obviously a very low-cost form of hunting.

Noctules have both a high aspect ratio and very high wing loading (flight speed 6.8 ms^{-1}). Even more extreme is Leisler's bat. Both bats are large, which will be a contributory factor to their high wing loading. However, even the noctule is no bigger than the serotine or greater horseshoe bat. This suggests that their high wing loading has evolved to be of some use to them, making noctules and Leisler's bats fast and efficient fliers, their efficiency increased by their high aspect ratio wings. The compromise is that they are best suited to open habitats since they have low manoeuvrability and their long wings could be a hindrance in dense vegetation. Noctules are indeed found in open habitat: foraging over water, pasture and the tops of trees. They are also known to forage over a wide area, often flying long distances each night. High flight efficiency is therefore important to them: they need to expend as little energy as possible in flight. Leisler's also forage over long distances and in open habitats. Given the differences in wing loading and aspect ratio of these two species, it would be interesting to compare flight performance and efficiency.

Wing shape

Of course, wing shape can vary in more subtle ways than can be described by aspect ratio. In particular the relative lengths and areas of the armwing and handwing can vary, and the wing tip can be rounded or pointed (Fig. 4.5). Variation in British bats tends to be rather subtle. With just two families, so few species, and relatively small differences in size and ecology amongst British bats, convincing differences can only be seen by looking at the extremes. The aerodynamics also gets more complex. The subtleties of how bat wings work are not yet understood, so we cannot be sure about the advantages and disadvantages of different designs. Often it is not even known whether a small dif-

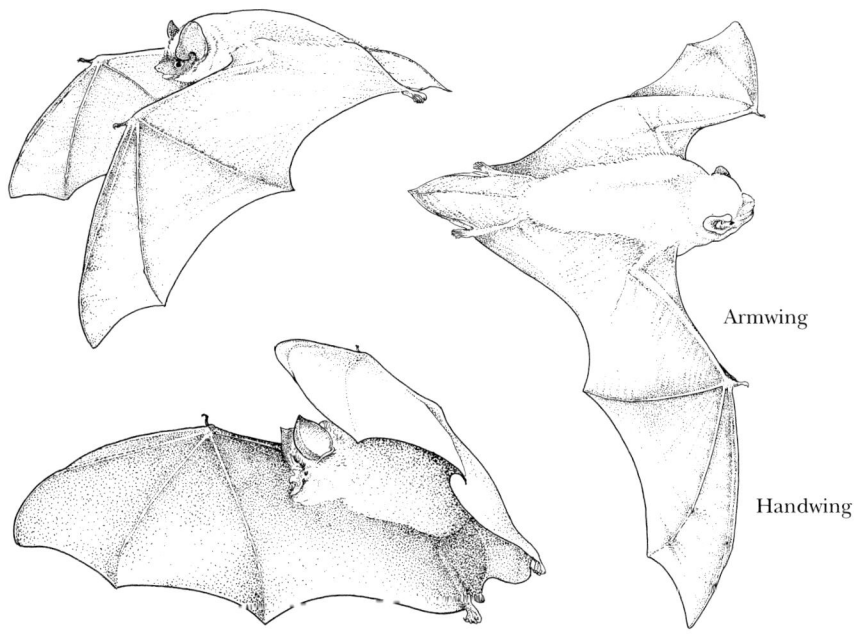

Fig. 4.5 Variation in wing shape of British bats.Clockwise from top: 45 pipistrelle, noctule and lesser horseshoe.

ference in some aspect of wing morphology is likely to lead to a difference in performance. However, there are some points that can be made with reasonable confidence. Bats with rounded wing tips are usually slow, manoeuvrable fliers. These species typically have short handwings. At the other extreme, fast-flying bats have long, pointed handwings. To explain this, the difference between manoeuvrability and agility needs to be understood. As discussed earlier, manoeuvrability is defined by the size of the turning circle. However, a bat can also change direction by going into a roll: tilting one wing sharply down initiates a turn down in the direction of the tilted wing. Bats with rapid rolls are said to be agile.

Fast-flying bats with long pointed wing tips, such as noctules, improve their rolling ability by reducing the inertia of their wing: inertia is the force resisting the acceleration of the wing in a new direction. They do this by having a relatively short armwing, thus keeping all the heavy bones near to the body. This may be demonstrated easily. Imagine your arms are wings, spread them out and move them rapidly up and down. Now put a heavy book in each hand and see how fast you can do it. Slow-flying species such as horseshoe, barbastelle and Bechstein's bats can also be agile, despite having short wing tips. These bats also have low aspect ratio wings, broad with long fingers. Broad handwings generate a great deal of lift and the long fingers allow them to curve the wing, or increase their camber, increasing lift further. This extra lift compensates for the high inertia. Remember that noctules must have narrow wings for other reasons, so this is not an option for them.

Power and flight

Now it is time to add another layer of complexity. Bats require considerable power to fly, and, whatever the shape of the wings, just how much power they need depends upon the speed at which they fly. Hovering, flying on the spot, is expensive: it requires a great deal of power and power has to be paid for, ultimately in food. As flight speed increases, the power required initially actually decreases. This is because the movement of the bat, or more precisely its wings, through the air reduces the power needed to produce lift and thrust. However, as flight speed increases further, there are dramatic increases in the drag on both body and wings, so the bat has to put more power into overcoming this drag. The relationship between the power required to fly and flight speed is therefore U-shaped, as shown in Figure 4.6. The exact shape is a source of some controversy amongst scientists studying it. Whatever the exact shape, power varies with speed and this has important ecological implications. If keeping down the cost of flight was the only important consideration, bats would always fly at one of two speeds. They would either fly at the speed that keeps the cost per second at a minimum – the *minimum power speed* – or at the speed that allows them to cover the maximum distance for the minimum cost – the *maximum range speed* (Fig. 4.6). However, the speed chosen depends upon the situation.

Consider, for example, a migrating bat that must fly from its summer roost to its winter hibernation site. It is not known for certain whether any of our bats migrate very far, but for this example, assume that they do. First of all we need to know whether the bat can feed during migration. If it cannot, then it has to fly as far as possible on the limited energy stores that it has. It therefore makes sense to fly at the maximum range speed. In this way it goes as far as it possibly can before it runs out of fuel. A bat might choose this option either because it has to make the journey quickly or because there is limited food available en route. If the bat can feed en route and is in no hurry, then minimum power speed may be more appropriate. What about a foraging bat? If food is plentiful, then minimum power speed may again be the most appropriate. In this way it uses the minimum amount of energy to stay airborne whilst it is foraging. However, at low insect densities it may not encounter food often enough at this slow speed, and may have to fly faster. To a commuting bat flying between roost and foraging site, the most important considerations may be getting there quickly and avoiding predators. Under these circumstances it may fly at speeds exceeding maximum range speed. Of course there are

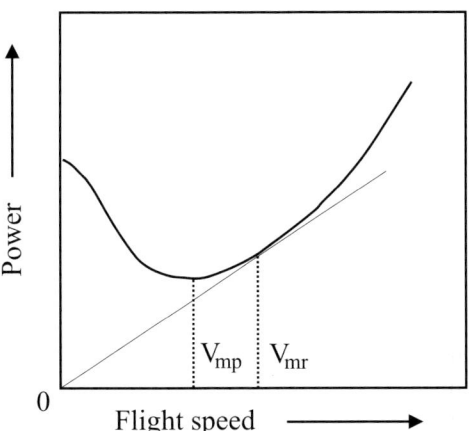

Fig. 4.6 The relationship between required power and flight speed. V_{mp} is the speed that requires minimum power, V_{mr} the speed giving maximum range.

many other factors that could confound this simple story, but work on pipistrelles has shown that foraging individuals do fly a little above maximum range speed, and commuting bats significantly faster. Daubenton's bats also fly faster than minimum power speed when foraging. Of course, optimal speeds are different for different bats. Each species will have its own power curve, to the left or right of the one shown, and above or below it. For example, relative to a greater horseshoe bat, the noctule is faster and has lower power requirements. Its power curve will be lower and to the right in Figure 4.6, if the curve shown is assumed to be that of the greater horseshoe bat. Unfortunately, as discussed earlier, it is difficult enough to get accurate measurement of flight speed in wild bats. Determining an entire power curve for a species is still a major technical challenge, but new technologies are making more and more things possible.

Power curves may even change within a species. For example, a pregnant bat will have a high wing loading and should have to fly faster, altering the power curve. Even a good meal can theoretically increase optimal speed. In theory, flight costs increase and the bat is less manoeuvrable. I doubt that a well-fed bat suffers a significant drop in performance, but pregnancy must have an effect. Oddly, captive pipistrelles studied through pregnancy and lactation decreased flight speed as they got heavier, incurring substantial increases in energetic costs since they flew at far from the optimum speeds predicted by theory. There are clearly lessons still to be learned. Female bats of many species are bigger than males and show other morphological differences, which may influence flight performance. For example, female Daubenton's bats are heavier than males, have a longer wingspan and a higher wing loading. Bats may even benefit from regulating their mass: there is no point in carrying excess weight that increases flight costs and decreases manoeuvrability unless the food supply is unpredictable or hibernation is imminent. Birds are known to regulate their mass in various ways. Small birds increase their fat stores late in the day to carry them through the night, rather than carry the excess weight right through the day. When food or weather are unpredictable they carry more weight as an insurance policy against this unpredictability. I see no reason why bats should not do the same.

There is an endless supply of other interesting and relevant aspects to flight, but two short examples will suffice here. Daubenton's bats feed very close to the water surface, rarely more than half a metre above it, and often only a few centimetres. As well as taking a significant proportion of their food from the water surface, they get another benefit from this behaviour. Changes in the air flow around the wings, due to the proximity of the water surface, can reduce the cost of flight if the bat is less than half a wingspan above the water: this phenomenon is known as ground effect. With a wingspan of 25 centimetres, Daubenton's bats should make significant energy savings, since they typically fly about 10 centimetres above the water. Did Daubenton's bats evolve their low-level flying style to exploit a new food source, or to reduce flight costs? Most probably both factors worked together in the evolution of this behaviour. Why are bats, particularly insectivorous bats, so small? Aerodynamic considerations are important here too. As bats get bigger their agility and manoeuvrability decrease. Insectivorous bats must be agile and manoeuvrable to catch their prey, so they need to be small. The task is made more difficult by their use of echolocation, a short-range detection system that rarely operates over more than a few metres. An echolocating bat has to react quickly to the presence of prey, since it has only a fraction of a second in which to alter its course and catch it. Echolocation brings with it yet another limiting factor. Most bats, at least during search phase flight, emit one echolocation pulse each wingbeat, usually late in the upstroke, which is when they breathe out. They breathe in on the downstroke. This coupling of the respiratory and locomotory movements of the thorax is believed to assist breathing, thus saving energy. However, due to mechanical constraints, bigger bats beat their wings more slowly and therefore emit echolocation calls less frequently. There comes a point when call emission frequency is just too low for a bat to effectively track and capture prey. This brings us to the next topic and another layer of complexity in bat ecology.

Echolocation ecology

Just as the shape of a bat's wings has ecological significance, so does the shape of its echolocation call. The last chapter described how echolocation calls take various forms and that each form can be shown to perform some tasks better than others. The calls used by a particular species should therefore have evolved to suit the habitat, prey and foraging strategy of that species. However, the patterns seen are not always easy to unravel. There is still a great deal not known about echolocation, compromises often have to be made, and there may be more than one solution to a particular echolocation problem. By working through the species considered in the last section, in the same sequence, we will see whether call structure may be linked to ecology.

The long-eared bats forage in and amongst vegetation, flying slowly, often hovering and often landing to take their prey. Moths are almost always the major component of their diet. Long-eared bats use low intensity, short, broadband FM echolocation calls, containing many harmonics (Fig. 4.7). A long-eared bat can fly within a metre or two of a bat detector and pass unrecorded.

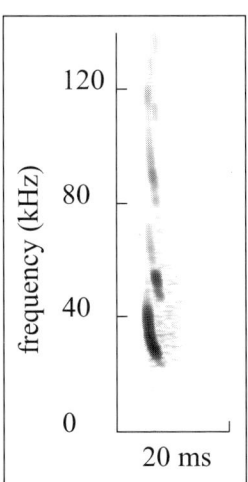

Fig. 4.7 Echolocation call of a brown long-eared bat.

To move safely in this cluttered habitat, echolocation must give long-eared bats a detailed map of their environment, and their broadband FM calls are ideally suited to this task. Short calls ensure that there is no overlap between a call and its echo from a nearby obstacle. This is important since the echolocation systems of most bats are intolerant of overlap. Very short, low intensity calls are also less likely to be detected by their prey: many moths and other insects have evolved 'ears' to detect approaching bats. In fact, long-eared bats will often stop echolocating in the final approach to their prey and locate them by passive hearing alone (p. 48). Their short, weak calls are not suited to detecting distant objects, but this is of limited importance to these slow-flying bats.

The next group of bats illustrates the point that rarely in biology is the picture straightforward. Natterer's, Bechstein's, barbastelle and lesser horseshoe bats all feed primarily in woodland. Their diets suggest that all but barbastelle bats feed by both hawking and gleaning, as non-flying and day-flying

Fig. 4.8 Echolocation
calls of Bechstein's,
Natterer's,
barbastelle and
lesser horseshoe
bats.

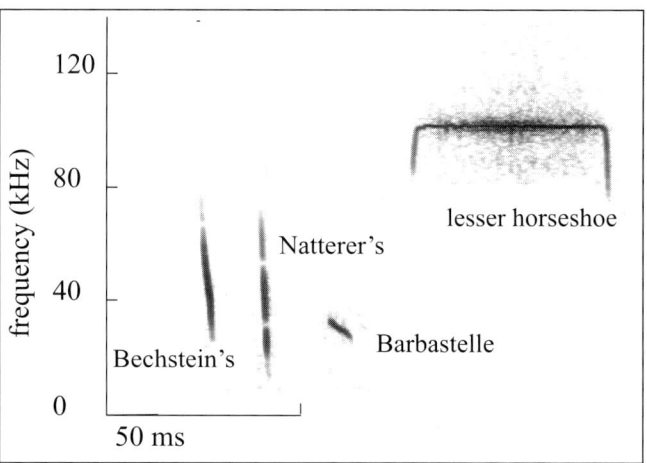

insects are taken and sometimes spiders. Barbastelle bats appear to feed mainly on moths taken in slow flight. Despite these differences, all these bats are relatively slow fliers that catch their food in woodland. However, there are major
differences in echolocation call structure (Fig. 4.8).

The two *Myotis* species, Natterer's and Bechstein's, have similar short, broadband FM calls, as might be expected. The barbastelle bat has a narrowband FM
call, that is sometimes a short inverted *j*, sweeping up, then down, the frequency range. The calls of the *Myotis* bats can be categorised as being broadband calls suitable for a cluttered habitat, and by stretching a point the barbastelle could be put in the same category. Why the barbastelle call has this
rather unusual structure is not known, but functionally it will work in a similar
way to the *Myotis* calls. But what about the lesser horseshoe bat? Its call is very
different: very high frequency, with a long central CF component at around
105 kHz and in most cases short FM sweeps at the beginning and end. The
long CF component of its call is ideal for detecting the fluttering wings of
insects, as described in the previous chapter (p. 45). The FM components presumably serve a similar function to the calls of the other species: orientation in
cluttered habitats. In most cases, the FM 'tail' at the end of the call has the
greater bandwidth (over 20 kHz), but the FM call at the beginning is often
prominent too and may be of equal importance. The greater horseshoe bat
has a similar call structure, presumably used in the same way, although the calls
are of lower frequency: the CF component is around 83 kHz. The large difference in frequency is largely due to the difference in size between the two
species. As bats get bigger, then on average their echolocation call frequency
decreases. In addition to catching food by aerial hawking and gleaning, the
greater horseshoe bat also flycatches from perches. The FM components are
present in calls emitted by bats hunting from perches, but increase in bandwidth when they are hunting in flight, suggesting greater importance. The FM
components in the calls of both horseshoe bats become more prominent during the feeding buzz, further emphasising their importance. It is easy to dis-
miss these FM components, since they represent such an apparently small part
of the call, but they are similar in duration and bandwidth to the exclusively
FM calls of many other species.

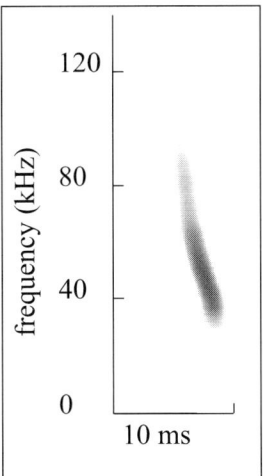

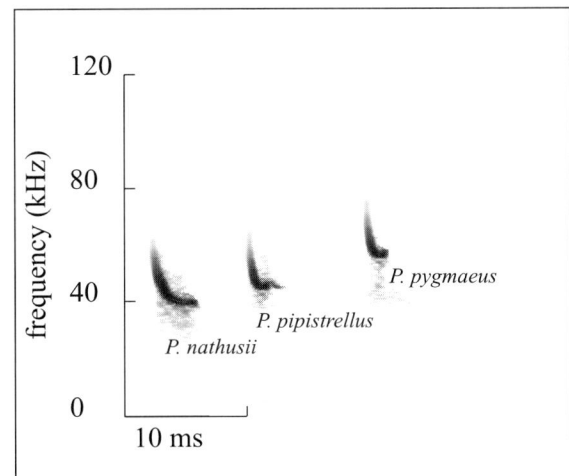

Fig. 4.9 Echolocation call
of Daubenton's bat.

Fig. 4.10 Echolocation calls of the three British
pipistrelle species.

Whiskered and Brandt's bats frequently feed in woodland too, and, interestingly, also appear to take a lot of non-flying prey. This is consistent with their wing morphology. Their short wings and low wing loading are suited to taking prey from vegetation by slow flight or gleaning. They also have echolocation calls characteristic of the other *Myotis* species that feed in woodland. Their extreme position on the graph in Figure 4.2, with very low wing loadings and aspect ratios, makes them worthy of further study.

Daubenton's bats have a wing morphology suited to their more open foraging habitat. However, their echolocation calls are almost indistinguishable from many of their woodland relatives (Fig. 4.9). This may be due to the fact that they catch much of their food from, or very close to, the water. Daubenton's bat echolocation calls therefore need to resolve insects on or close to the water surface: a demanding task.

The three pipistrelle species all forage in more open environments than any of the species discussed so far. When foraging in woodland they confine themselves to rides, clearings or edges, although they will feed within a metre of obstacles. Typical calls are shown in Figure 4.10. These species probably have the most variable calls of any of our native bats. When flying in the open, all use narrowband calls, with a bandwidth of less than 15 kHz. In more cluttered habitats this can increase to about 60 kHz. At the same time, the duration of the pulse decreases, from 5 milliseconds or more to 3 milliseconds or less. This makes sense for several reasons. Broadband calls are better at resolving clutter, so bats flying nearer vegetation would be expected to increase bandwidth if that is possible. A shorter duration avoids overlap between the call and the echo from nearby obstacles. Assuming the calls have a similar energy content in the two situations, then in the narrowband calls there will be more energy at a given frequency. The echoes from these calls from distant objects are therefore more likely to be audible to the bats than those from broadband calls. This may increase the range over which the bats can detect prey in the open. In fact, even when emitting broad-band calls, much of the energy is concentrated in the narrowband tail of the call in all but the most extreme FM calls.

The narrowband signal varies considerably in the pipistrelles, with peak frequencies (frequency of maximum energy) of 39, 46 and 55 kHz respectively. Although there is considerable variation within each species, there is little overlap between them. Theory suggests that as the frequency of the call increases, smaller insects can be detected, but there is no evidence to suggest that these bats take different-sized prey. The higher frequency calls attenuate more rapidly in air, so Nathusius' pipistrelle should be able to detect prey at greater distances than the 55 pipistrelle.

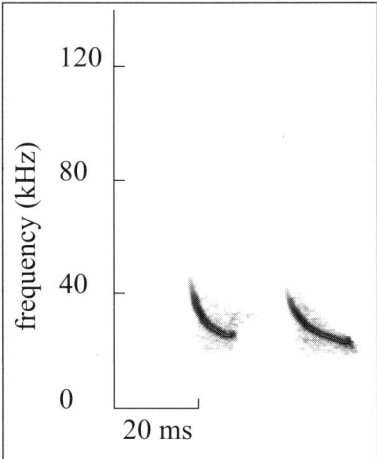

Fig. 4.11 Echolocation call of the serotine.

Returning to Figure 4.2 and moving right on the diagram, the next bats are the greater horseshoe bat and the serotine. The greater horseshoe bat has already been discussed. The call of the serotine is shown in Figure 4.11. It too is variable, showing similar changes to those of the pipistrelles, and it is very likely that call structure changes for the same reasons. The serotine has a peak frequency of 32 kHz, although it can vary between 26 and 42 kHz. Since these values are from calls recorded from bats close enough to be reliably identified, or released from the hand, they may not be fully representative. Individuals flying higher and further from clutter may have calls of even lower frequency. The lower frequency call will be advantageous to this larger, faster-flying and therefore less manoeuvrable species, since prey can be detected at greater distances.

Finally, there are two related species with very high aspect ratios and high wing loadings: noctule and Leisler's bats, fast and efficient fliers exploiting open habitats. Both feed by aerial hawking, but surprisingly, the noctule will occasionally take insects from the ground. Echolocation calls of the noctule and Leisler's bat are shown in Figure 4.12. Both have high intensity, long, low frequency, narrowband calls, ideally suited to their foraging style. The calls are resistant to attenuation, and yield strong echoes over a limited frequency range. Calls increase in bandwidth and decrease in duration on approaching clutter. Both noctules and Leisler's bats frequently alternate broadband and narrowband calls in flight, getting the best of both types of call. The long calls may even be effective flutter detectors. Is there a downside to this type of call? Low frequency calls should not be good at detecting small insects, yet both species will feed on small insects. The theory predicting this is based on echoes from simple targets such as spheres and disks. There is evidence to suggest that echo intensity from the more complex surfaces of insects is less dependent on echolocation call frequency than might be supposed.

This introduces another aspect of call design not so far discussed, the 'duty cycle'. This is the length of an echolocation call relative to the interval between calls. Most bats have short duty cycles: short calls with long gaps between them.

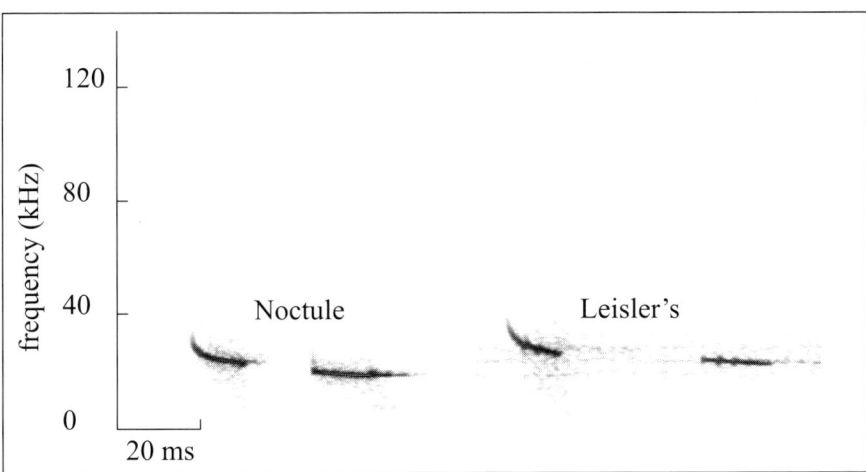

Fig. 4.12 Echolocation calls of noctule and Leisler's bats.

The primary reason for this appears to be the avoidance of pulse-echo overlap, since the echo appears to be masked by the outgoing call.

As bats get bigger and their average call frequency decreases, their calls tend to get longer. This is shown in Figure 4.13 for bats that use FM sweeps with a pseudo-CF tail. It would not make sense to compare bats that use very different call types. There are not enough of these bats in Britain to make such a comparison, so other species from around the world are included in the graph. Why do calls get longer as frequency decreases? As call frequency decreases, atmospheric attenuation is less. This means prey are detected at greater distances and pulses can be longer without the problem of pulse-echo overlap. This is particularly important for larger bats because they are usually

Fig. 4.13 Call duration in relation to call frequency.

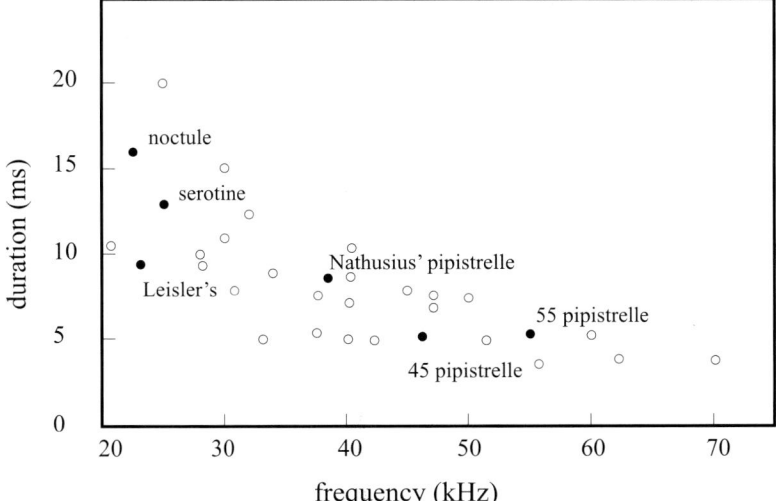

faster and less manoeuvrable than smaller bats and late detection would reduce their ability to catch prey. This should mean a decreased ability to detect small prey, but, as discussed above, this does not appear to be the case. Horseshoe bats are unusual in that they use calls of high frequency and high duty cycle. They can do this because they are not susceptible to the problems of pulse-echo overlap. This is due to the Doppler shift compensation mechanism described in Chapter 3, which separates pulse and echo by frequency, allowing them to be processed at the same time.

Summary

Within the vesper bats, species that feed in cluttered environments have low wing loading and low aspect ratios for slow, manoeuvrable flight. To orientate themselves in this complex habitat and detect their prey, they have short, broadband echolocation calls. Species that fly in more open habitats tend to have higher wing loadings and aspect ratios that confer faster, more efficient flight, but make these species less manoeuvrable. These bats have longer, narrowband calls for the detection of more distant targets. Larger, faster and less manoeuvrable bats have the longest calls with the lowest frequencies, but this does not appear to restrict them to large prey as theory predicts. Horseshoe bats have wings typical of other bats that feed within or near clutter, but have evolved a very different echolocation system, which has some distinct advantages.

Feeding ecology

Flight and echolocation equip bats to feed on a wide range of insects, as well as other invertebrate groups such as spiders. Variations in wing shape and echolocation call structure favour different foraging strategies, habitats and even diets. It is not surprising then that there are dietary differences between species. However, differences in diet may reflect more than just differences in where bats feed. Several factors might contribute to differences in what insects are eaten, relative to those available in a particular habitat. For a bat of a given species, some insects may be too small to detect, some may be physically too large to catch and hold. Others may have strong exoskeletons and be too tough to eat. Some may have escape mechanisms or particular behaviour patterns that make them too difficult to catch. More subtly, the diet may not only be unrepresentative of the potential prey present, but also of those that *can* be eaten. For example, some insects may be not worth the trouble of catching and eating: the energy expended in catching and eating them may be greater than the energy obtained from them. This could be because they are very small, or because they are large and difficult to carry, chew or swallow. Some insects are unpalatable or even toxic. To resolve some of these issues, specific questions need to be answered.

A good starting point is to ask whether the composition of the diet directly reflects the relative abundance of the insects available to a species. If, for example, mayflies make up 10 per cent of the insects flying over a river, and they make up 10 per cent of the diet of Daubenton's bats, then this suggests the bats neither avoid, nor specifically seek out, mayflies. If, on the other hand, they make up 50 per cent of Daubenton's bats' diet, the question is why do they eat so many mayflies? Do the bats choose mayflies in preference to other insects because they are very nutritious, or are they for some reason easy to catch?

Posing these questions is easy enough, answering them is rather more difficult, and to date few studies have attempted it. However, some information is available for some species. Before looking at the evidence, a note of caution. Dietary estimation is subject to many biases, most of which are hard to overcome or even evaluate. The insects available to a bat are assessed by various trapping methods, many of which are selective. Not all insects are attracted to light traps; large, strong fliers may not be caught by suction traps; small insects may pass through some nets; it is not always possible to sample insects in the precise microhabitat used by the bats. Assessing what the bat eats is also prone to error. Some insects are not well preserved in faeces; transit time through the gut may vary; some are simply hard to identify or size; some are dismembered before being eaten. Finally there are many ways of expressing the results, each with its own bias: by number; by volume or mass; by simple presence in a dropping. Bearing these difficulties in mind the evidence may be presented.

One of the most detailed studies, on the greater horseshoe bat, clearly shows that these bats can be selective. These details may be unique to this one population of bats at the time of the study, but the basic pattern can probably be applied to greater horseshoe bats in general. Diet was assessed by analysis of droppings and by insect remains under feeding perches. First, greater horseshoe bats eat more large prey than expected: 50 per cent of the insects eaten had a wing length greater than 15 millimetres, yet insects of this size comprised only about 5 per cent of those captured in a light trap. The bats ate few insects with wing lengths less than 4 millimetres, which is the wavelength of the CF component of their echolocation calls. This suggests that smaller insects may not be easily detected. However, it is also possible that it is more profitable to ignore them and catch larger prey. It takes a few seconds to catch and eat even a small insect, and maximum capture rates have been measured at around ten insects per minute. So for any bat there must come a point when an insect is simply too small to contain sufficient energy to make it worth catching: better to save time and energy and find a larger insect. Very large insects are usually taken to a perch before being dismembered and eaten. This also makes energetic sense. Bigger insects take longer to eat: they have longer handling times. Since flying is around seven times more expensive than hanging from a perch, if the insect is going to take some time to eat, energy can be saved by landing to eat it.

Greater horseshoe bats eat mainly moths and beetles. By volume, 40 per cent of the diet in this study was moths and 30 per cent beetles. However, they eat a wide range of other insects, and both the composition and the diversity of the diet can change with the seasons (Fig. 4.14).

Early in the year, in April and May, craneflies, dor beetles and cockchafers were the most commonly taken insects. The volume of moths eaten increased dramatically through May to a peak in June, before falling slowly throughout the summer and autumn. Moths were steadily replaced by dung beetles from July through to October. The diversity of the diet was greatest in spring and autumn, and lowest in June when it was almost exclusively moths, and July when it was moths and beetles. There are several possible reasons for this change in diversity. The balance of the evidence suggests that as more moths become available, more are eaten, in preference to other insects such as flies. The diet is probably diverse in spring and autumn because the bats' preferred food, moths, is scarce. The well-defined periods when craneflies and various

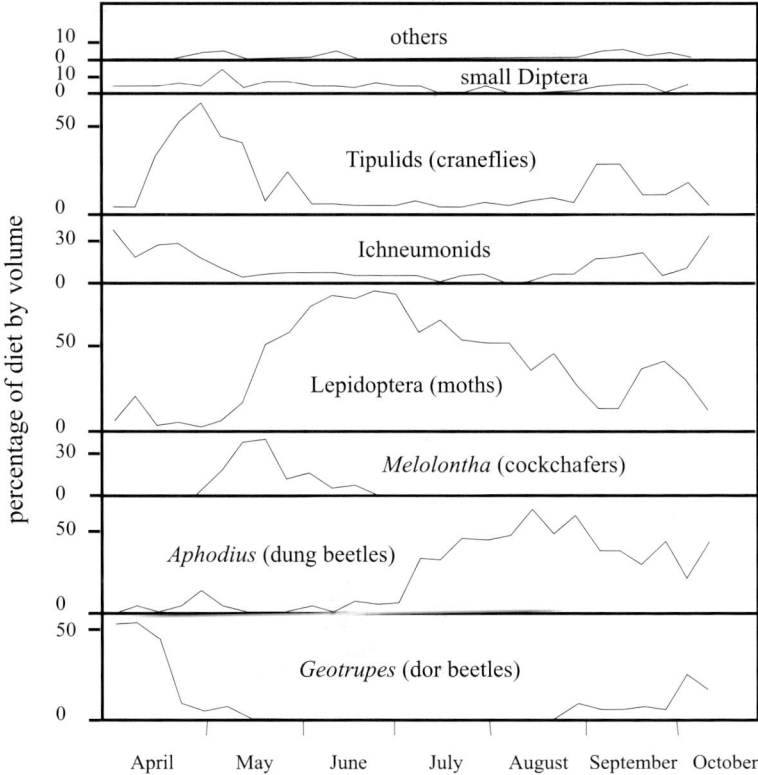

Fig. 4.14 Seasonal changes in the diet of the greater horseshoe bat (from Jones, 1990).

beetles were major components of the diet may be related to their seasonal abundance. At different times, these groups were 30–50 per cent of all the insects taken, a proportion that greatly exceeded their relative abundance in the field. Are they taken in such large numbers simply because they are so large and conspicuous, or do the bats actively seek them out in preference to other insects? As in all such studies, there are other factors, difficult or impossible to control, that may contribute to the change in diet over the summer. Whilst interesting in themselves, they unfortunately make it difficult to determine which mechanisms are operating when several are available. Increased wing loading during pregnancy is known to alter flight performance and that may make some insects harder to catch. The bats also changed roost site late in pregnancy and that may have led to a change in available prey. Finally the droppings analysed came initially from only adult females, but later in the year droppings were also from their offspring, which may have had a different diet.

Other studies also show that diet is not always a simple reflection of the relative abundance of the food available. For example, pipistrelles in one study were shown to select for mayflies and against moths. Pipistrelles probably avoid moths because most of them will be too large to handle, but why did they select mayflies? One possible reason is that they may have been present in large swarms, making them conspicuous and easy to catch.

Can any simple conclusions be drawn? It appears that most species, most of the time, are relatively unselective, taking whatever insects are available to them roughly in proportion to their abundance in the environment in which the bats feed. Availability is probably determined by what can be detected, caught and eaten. This naturally constrains some small bats to smaller prey and some large bats to prey above a certain minimum size. Some of the time, some species are selective, perhaps favouring prey of a certain size. Some species appear to select particular insects most of the time, most notably moths. Some of those specialising in moths have evolved particular strategies to catch them. These will be discussed in more detail later in the chapter (p. 97).

Roosting ecology

The last chapter considered those aspects of roosting most relevant to hibernation and reproduction. This section studies roosting in a broader context, filling in some of the gaps, starting with a list of those factors that influence the choice of roost and the way in which bats use their roosts.

Most importantly, roosts provide bats with shelter from the extreme conditions imposed by the outside world, principally weather and predators. They can also provide an appropriate microclimate for the bats' current physiological needs. Communal roosting potentially improves mating opportunity, maternal care and the transfer of useful information. If there is some choice in the location of the roost, it can reduce commuting distance, minimising the energetic and predation costs of foraging. Roost choice may frequently require compromise, since a single roost may not fulfil all these requirements. Roost choice may indeed be limited and have a direct effect on the distribution of bats, independent of feeding requirements.

Roost choice and microclimate

Bats are small mammals and as such have a large surface area in relation to their mass. Their wings increase surface area enormously relative to other small mammals. Simple physics shows that small bodies generate little heat and lose this heat quickly through their surface. Insulation helps, but only large mammals can rely on it: it would take a ball of fur as big as a grapefruit to keep a pipistrelle warm, but this would compromise other activities! Choosing an appropriate roost site is therefore an important part of their thermoregulatory strategy: one with low, stable temperatures and high humidity for hibernation, a warm roost in which to bring up their pups. Individual bats, on a day-to-day basis, may have varying requirements and move around within a roost or between roosts. Because most temperate bats are heterotherms they can regulate their energetic requirements by regulating body temperature. The easiest way to do this is to choose a roosting site with the appropriate ambient temperature. Bats can alter the temperature of their roost, by roosting in large groups, and by using small cavities that retain heat. A blind cavity in the roof of a cave or building traps effectively the rising heat given off by the bats themselves.

Natural roosts in Britain are tree holes, caves and the many minor cavities in exposed rock surfaces. Caves were discussed in the previous chapter under hibernation (p. 51), but they are occasionally used in the summer months, although rarely as breeding sites. Small groups of bats use caves when feeding conditions are poor and cool sites are needed to save energy. Tree holes must be the most important natural summer roost sites. Before humans destroyed most of the natural forests, and 'tidied' those that remained by removing old and decaying trees, natural roosts would have been abundant. Even now, trees may well be the most important roosts for all British species except pipistrelles, serotines and horseshoe bats. Roosts range from small cavities under flaking bark that house no more than a single bat, through holes and cracks in branches to large hollow trunks. Medium-sized cracks in high limbs are probably the most abundant and the most useful to bats. Finding such roosts is not easy, and until recently most were discovered by accident. In recent years, the development of small radio transmitters has made it possible to track even the smallest species, leading to the discovery of more and more tree roosts. This work further highlights the difficulty of finding such roosts: having found the tree, it can still be a major task locating the roost itself. Wood is a good insulator and many cracks offer protection against both predators and climatic extremes. Tree holes shaded by the canopy are relatively cool, but sunlit holes in dead trees can be warm and may be favoured as nursery roosts. Rock crevices have been largely overlooked, but are probably important in those parts of the country where they are common: Daubenton's bat roosts have been found in the limestone scars of the Yorkshire Dales. These too offer a range of microclimates: horizontal crevices in south-facing crags can be surprisingly warm, vertical cracks cool quickly and deeper cracks are more effective at buffering temperature extremes. Bats can move around in such cracks to find a suitable microclimate as the weather changes.

Buildings offer an enormous range of microclimates, and those bats that use buildings have learned to exploit the many possibilities open to them (Fig. 4.15). The roof space of a building may be hot when heated by the sun and the

Fig. 4.15 Roost sites in buildings.

temperature gradients within it can be substantial. Internal heating and the presence or absence of insulation increase the range of microclimates available. Cavity walls, basements, window frames, shingle and weather-boarded walls all provide additional opportunities: there are few cavities within buildings that have not been exploited by bats. Crevices in the timber parts of buildings can be thought of as surrogate tree holes and gaps in the stone or brickwork of buildings and bridges seen as substitutes for rock crevices. In the same way, mines, tunnels and other underground structures are obvious roost choices since they share many characteristics with caves.

A number of attempts have been made to analyse what it is about a building's structure that makes it a good bat roost, but few general patterns have emerged and a discussion of this topic is best left to individual species accounts (pp.109–135).

Roosting and foraging

Roost choice and habitat

In addition to having the right microclimate inside them, the best roosts will also be close to good foraging habitat, preferably with safe commuting routes between the two. For most species this means low altitude sites close to woodland and water. Pipistrelles, for example, are much more likely to use a building if it has even a small number of large trees near it. Significant woodland cover, with hedges to link woodland to the building, makes a building a more desirable residence. For many species, the presence of freshwater habitats, flowing or static, increase the value further.

The characteristics that make good natural roosts, primarily those of trees, have received little attention in Britain. Daubenton's bats in the Yorkshire Dales show a strong preference for holes in ash trees. This may be because ash is the most common naturally occurring species along the rivers, and cracks and cavities readily form even in quite young trees. Although other tree species are abundant, some do not readily form cracks. Introduced deciduous trees, even if cavity-forming, may not have been around long enough to have been adopted by Daubenton's bats, although this seems unlikely given the speed at which they have taken to buildings and bridges. In other parts of the country, other bats appear to favour different trees: no clear pattern has yet emerged, but oak, ash and alder appear to be used more frequently than other species. In a

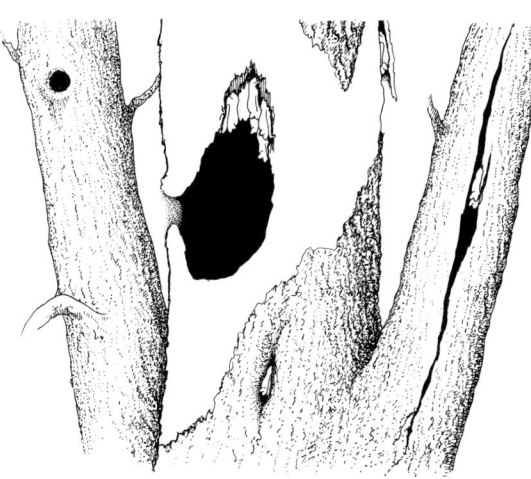

Fig. 4.16 Natural tree roosts.

study in the Netherlands both noctules and Daubenton's bats preferred oak to beech and both preferred trees on the edge of woodland. Interestingly, noctules had a marked preference for woodpecker holes over natural cavities, suggesting a significant dependence on the presence of woodpeckers.

Roost fidelity

Few colonies use a single roost site, and most have many roosts within their home range, although only a few may be in use at any one time. Colonies frequently break up, using two or more roosts at the same time. Indeed it is difficult to define a colony in many instances, due to the constant fragmentation, regrouping and movement of roosting groups. This question will be discussed further in the next section. For the moment a colony may be defined as a group of bats that frequently roost together. There appear to be few hard and fast rules when it comes to roost fidelity. Most colonies appear to have a small number of favoured roost sites, which are used over many years. In extreme cases a coherent colony may use a single site for many decades. Others will move in a largely predictable way between a small number of roosts, often on a seasonal basis, with perhaps one roost used in the spring, another during pregnancy and a third to give birth and rear the pups. At the other extreme many roosts may be used in an apparently random fashion, perhaps with frequent changes in composition of the colony subgroups.

What are the major driving forces for these movements? If the discussion is restricted to summer nursery colonies, four factors come to mind: roost microclimate, foraging success, predation pressure and parasite burden. Microclimate was discussed earlier (p. 87), but recent work has shown how very important it is in determining roost choice. Work on Bechstein's bats in Europe, using bat boxes, has shown that roost temperature is the most important factor in determining whether or not a group of bats will change roost. Predation and parasites will be discussed at the end of the chapter (p. 100), so this section deals with foraging success.

Roosts and foraging behaviour

I have show how echolocation and flight morphology determine a bat's foraging strategy and habitat, and discussed flight speed and efficiency, but how far will bats fly for their food? Because flight is a fast and efficient mode of transport, bats are able to cover large areas in search of food. However, there are no advantages to flying further than necessary, since it wastes energy, reduces foraging time and increases predation risk. Bats might therefore be expected to roost close to good foraging sites whenever possible. But how close is close? How large is a bat's home range? As would be predicted, the average distance a bat flies between roost and foraging sites is to a large extent determined by its flight speed and efficiency: large, fast and efficient aerial hawkers will fly further than small, slow, gleaning species. However, there are always exceptions. Even within a species, distances are so variable that a large number of bats must be studied to get a true picture of their behaviour. For example, individual Daubenton's bats have been radio-tracked one at a time, for 1–2 weeks each, as they foraged on a river in the Yorkshire Dales. The roosting and foraging patterns of a few typical individuals are summarised in Figure 4.17. Roosts are shown by circles and foraging sites by squares. Ellipses enclose the roosts and foraging sites of individual bats over the 1–2 weeks each bat was

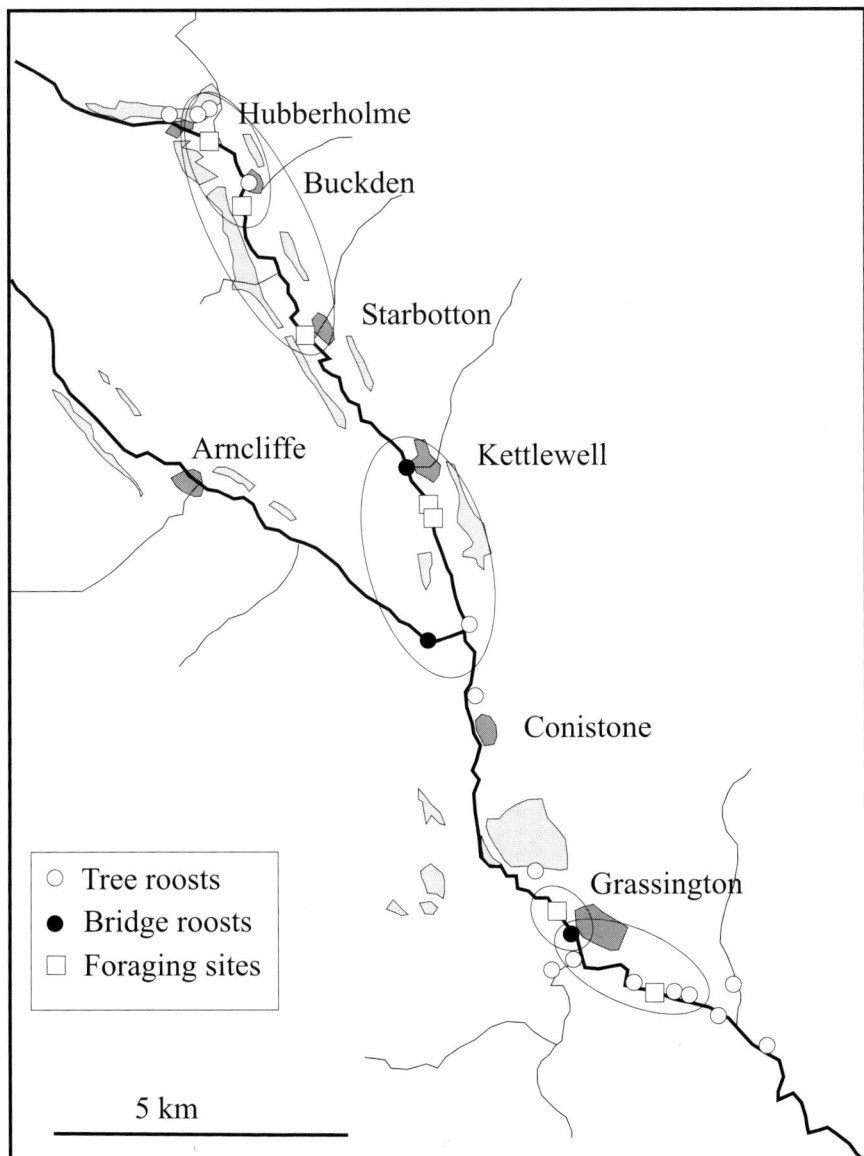

Fig. 4.17 Map of Daubenton's bat movements in Wharfedale, Yorkshire Dales.

tracked. Over the entire tracking period, some bats never flew more than 750 metres from their primary roost, while others flew up to 16.4 kilometres. The average distance travelled was between 1 and 6 kilometres, depending upon the roost being studied, and nursing females typically foraged only 0.5–2 kilometres from their current roost site. However, these bats frequently changed roosts. Roosts were in the stonework of bridges, occasionally in natural crags, but most frequently in cracks and holes in river bank ash trees: the dominant native tree in the Dales. The bats' longest flights were often made on those occasions when they changed roost. In some cases the pattern of behaviour suggested that the roost switch was made to enable bats to forage at a new location close to the new roost. For example, one female fed over several nights at a site less than 1 kilometre downstream from its roost. It left this site late one night to feed 2 kilometres further downstream and spent the night in a tree roost less than 200 metres from this new foraging site. For the next few nights it used both the new roost and the new feeding site. Was the bat taking advantage of a better feeding site or being forced to switch roost (and therefore feeding site) for another reason? A male bat similarly fed every night for 13 nights at a still pool in the river 500 metres from its primary roost. It spent two of its 13 nights over 2 kilometres downstream at a second roost, but still commuted to its original foraging site each night. Clearly, minimising commuting distance was not the reason for the roost switch in this case.

Home range in other species has also been determined by radio-tracking. In most cases, home range can be expressed in terms of commuting distances or foraging areas, since the bats radiate out from their roost site(s) each night. A study in Kent and Avon showed that Leisler's bats flew up to 5.8 kilometres from the roost (mean maximum distance 4.2 kilometres). They had an average home range area of 7.4 square kilometres, but it could be as large as 18.4 square kilometres. In a similar study in Ireland Leisler's bats flew up to 13.4 kilometres from the roost (mean maximum distances 4–7 kilometres). In both studies, the bats made use of several alternative day roosts in buildings and trees. Serotine bats travel over similar distances, although in one study home range areas of up to 48 square kilometres were recorded. The noctule has not been extensively studied, but it probably travels similar distances to feed. The only published data, from Europe, records a maximum distance of only 2.4 kilometres, which I suspect is not typical. Greater horseshoe bats, although large, typically forage only 1–3 kilometres from their roost, but occasionally travel 10 kilometres. Similar results have been obtained for several other species. Pipistrelles usually feed within 3 kilometres of the roost, brown long-eared bats rarely forage more than 1 kilometre from their roost. Recently, the rare Bechstein's and barbastelle bats have also been studied. Surprisingly, in a study in Sussex, barbastelles ranged as far as 4.5–18 kilometres from their roosts and some individuals had home ranges greater than 20 square kilometres. Radio-tracked barbastelles in Somerset also appeared to have large home ranges. Bechstein's bats, on the other hand, like long-eared bats, forage within 1 kilometre of their woodland roost and have individual home ranges of just 0.1–0.5 square kilometres.

Territoriality

Territoriality may be an important factor in the behaviour and distribution of foraging bats. However, there have been few specific attempts to study such behaviour, perhaps due to the difficulties involved. The most obvious sign of territoriality is aggression, although absence (or apparent absence) of aggression cannot be taken as evidence for a lack of territoriality, since a territory may be held with only occasional shows of aggression. Aggressive behaviour usually manifests itself in chasing behaviour and vocalisation, but both can be used in non-aggressive circumstances. Chasing, accompanied by audible vocalisation, is used by foraging pipistrelles to chase away other pipistrelles straying into their foraging area. This behaviour is generally rare, being seen only at low insect densities, suggesting that the bats become territorial only when food is scarce. Similar behaviour has been seen in Daubenton's bats and several North American species. Foraging Daubenton's bats typically distribute themselves along a stretch of river and each bat has a short, well-defined beat. In travelling down the River Wharfe in the Yorkshire Dales the number of bats on a stretch of river increases. Each bat occupies a smaller beat and chasing becomes more frequent, but the average feeding buzz rate, per bat, remains constant. Are the bats defending a foraging beat big enough to sustain an optimal feeding rate? An alternative explanation is that they are simply distributing themselves non-aggressively so as to maximise their feeding rate.

I recently watched two unidentified *Myotis* bats (probably whiskered or Brandt's) flying for over 30 minutes outside a cave from which they had just emerged. Feeding buzzes were heard frequently, but the two bats spent at least half their time chasing each other. On one occasion the bats met and tumbled, rotating together for a metre before continuing the chase. It was early April, two hours after dark and the temperature was barely above freezing. Pipistrelles that had been feeding earlier in the evening had left the site, possibly due to a lack of flying prey. This is an energetically expensive behaviour for bats recently emerged from hibernation and probably hungry, in less than perfect foraging conditions. Was this territorial behaviour or did it serve some other function?

Information transfer and group foraging

A potential benefit of both communal roosting and foraging is the transfer of useful information, such as the location of good feeding sites or alternative roost sites. This transfer of information may not always be very sophisticated, nor even deliberate. In a study in the United States, evening bats, *Nycticeus humeralis,* marked with colour-coded rings were weighed and photographed automatically as they entered and left the roost. Bats that had not fed well on a previous foraging trip followed colony members as they left on their next foraging trip. In doing so, they increased their chances of getting a good meal. There was no evidence in this study that bats could recognise their well-fed roost mates: it seems that by randomly following other bats they increased their chance of feeding success. This behaviour may occur in some of our own bats and may be one reason why bats often emerge from the roost in clusters rather than as a steady stream. Groups of both commuting and foraging bats are observed frequently. Pipistrelles marked with coloured reflective disks have been shown to leave, forage and return to the roost in small groups, at least some of the time. Individuals in these groups arrived at a foraging site, fed on separate beats and later moved on as a group. Group foraging may increase the chances of finding rich food patches. Most insects, particularly larger ones, are not uniformly distributed through either habitat or time. Insect distribution will be determined by season, habitat and microclimate, by emergence, swarming and feeding behaviours and no doubt other factors too. In many instances, this will lead to very patchy insect distributions. The first task of a foraging bat is to find the rich patches. By foraging in loose groups, patches may be more easily found since the group will cover more ground than a single individual. Evidence for information transfer might be expected to be found most easily in gleaning bats, but recent work on Bechstein's bat in Europe found no evidence for group foraging or information exchange. Individual bats were faithful to particular foraging sites that were often some distance from each other. The benefits of communal roosting are probably the important driving forces, although familiarity with local patterns of insect distribution, flight lines, night roosts and resident predators will all lead to more efficient and safer foraging.

Having found a patch of insects, the feeding buzzes of one individual may attract other bats in the vicinity. The feeding buzz of many bats shows a marked drop in frequency in the terminal phase. Figure 4.18 shows this terminal drop as a Daubenton's bat approaches its prey. It is not obvious what function this terminal frequency drop may serve in terms of target discrimination. If anything, it should compromise performance. However, this low frequency component travels further than other parts of the call. It has been suggested that it may serve as a signal to other bats, attracting them to the insect swarm. In some circumstances this could simply increase competition for food, but in others it may increase individual foraging success, by reducing the effectiveness of any countermeasures the prey are able to take. Can this low frequency component be switched on and off on demand? This is not known. Bats of many species emit a variety of social calls when feeding. Some may be for territorial defence, others may be calling bats to a food source. As yet, very little is known about this aspect of behaviour in British bats, but there are clues from studies in other parts of the world. The spear-nosed bat, *Phyllostomus hastatus*, in South America, and the Australian ghost bat, *Macroderma gigas*, to name just two, emit audible social calls when feeding, with no evidence of territorial behaviour.

In many cases, perhaps most, commuting and foraging are distinctly separate behaviours. Most bats appear to leave the roost and fly rapidly and more or less directly to favoured foraging sites. Commuting routes frequently (but not always) follow obvious features in the landscape, such as hedgerows, woodland edges, lanes, waterways and even small ditches and dykes. Bats frequently fly close to these features. This behaviour may in part be an anti-predator strategy, but it may also facilitate navigation.

On returning to the roost bats rarely enter directly, but circle the entrance repeatedly. Several false landings may be made before the bats finally enter the

Fig. 4.18 Feeding buzz of Daubenton's bat showing a drop in frequency just before prey capture. (Note the apparent double pulses early in the feeding buzz – the second pulse of each pair is in fact a reflection of the real pulse off the smooth water surface, which arrives at the microphone just after the real pulse due to its longer path length.)

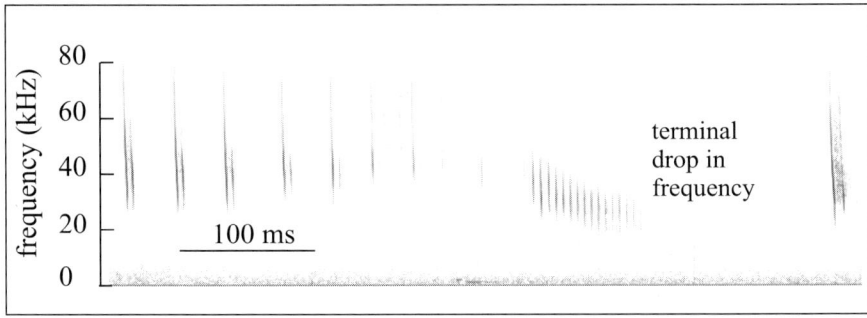

roost. Large and conspicuous swarms may form during the hour before sunrise and social calls are frequently heard. Just why the bats do this is a mystery, but there must be a very good reason, since the bats use a great deal of energy flying in circles and greatly increase the risk of being caught by a predator. None of the reasons I can think of to explain this behaviour, nor those I've heard suggested, are convincing. Are the bats just too clumsy to get into the roost first time? Do they need constantly to remind themselves of the structure and location of the roost entrance? Does circling perform some social function?

Population structure

Until recently most of what was known about the structure of bat populations had come from the slow process of ringing bats in the hope of recapturing them at a later date. In recent years, interest in the structure of bat populations has grown considerably. This has been fuelled by technical advances in genetics and by concerns about the conservation of small, isolated populations of some bat species as well as by scientific curiosity. It is now possible to answer questions about the maternity or paternity of individual bats, relationships within and between colonies, between species, and about the evolution of bats. All this can be done by DNA analysis using a tiny fragment of wing membrane that can be removed painlessly and easily, leaving a small hole that rapidly heals. This molecular approach, in combination with ringing, radio-tracking and other techniques, has enormous potential for revealing the detailed genetic and social structure of bat populations.

Early work has shown that nursery colonies are usually made up of females that return year after year to the same roost or group of roosts. Female offspring usually return to the same colony after hibernation, whereas males frequently disperse. It was assumed that inbreeding was minimised and essential genetic diversity was maintained through male dispersion. Recent work supports this basic idea, but shows that the picture is rather more complicated. Several females in a study colony of greater horseshoe bats mated with the same males over several years, and males often fathered offspring in the colony in which they themselves were born. Despite this, neighbouring colonies show distinct genetic differences, probably due to mating between bats from more distant colonies and some permanent migration of bats between colonies.

Similar results were obtained from a study on brown long-eared bats: a small but significant proportion of the offspring were fathered by males born in the same colony. But again, there is probably extensive mating with males from other colonies, maintaining genetic diversity. It is also probable that there are several or even many maternal lines within a single colony, so that breeding between males and females from the same roost need not mean breeding with close relatives. Nursery colonies of mouse-eared bats in Europe are also made up predominantly of females, although small numbers of males are sometimes present. Again, ringing studies showed that males rarely return to their natal roost after hibernation, but the majority of the females do. In a study of a particularly large roost of 700 females in Germany, there were 19 resident males. The resident males were not responsible for fathering many of the offspring in the roost, which is not surprising, since the females are known to leave the roost in search of mates, visiting roosts up to 12 kilometres away. What was surprising was the observation that the males sought by the females were related to them: probably dispersed offspring from their own colony.

Bat communities

In earlier sections, some of the advantages and disadvantages of group living within species were discussed. Few bat species have their habitat or ecological niche to themselves, but must share them, at least to some extent, with other bats. At any particular location it is possible to record or capture several or even many species of bat: it is rarely the preserve of a single species. To coexist, different species must be able to share resources, to exploit them in different ways, or to use different resources to avoid direct competition. Our landscape is a mosaic of ecological niches. A single location, such as a river, wood or even a single tree, can have several niches, separated from others in space, in time, or by the behaviour or food preferences of the animals within it. Whilst bats eat the insects in a wood by night, birds and other animals eat them by day. Ecological niches, like habitats, are often not easily defined, and may show considerable overlap. All our bats are insectivorous, and most depend to a large extent on the most abundant insect groups, such as flies. The majority of our bats also roost in tree holes. To what extent are our bats in competition with each other for resources? Is there major overlap of ecological niches, or have bat communities evolved to minimise competition? The beginning of this chapter described how wing form and echolocation call structure have evolved to equip bats for different foraging strategies, but each species was discussed in isolation. This section looks explicitly at the question of coexistence.

Set up a 3-metre high, 2-metre wide harp trap (p. 172) in a small gap between trees, on a typical summer night, directly over a narrow stretch of river in the north of England. In addition to catching Daubenton's bats, you may be lucky enough to catch 45 and 55 kHz pipistrelles, Natterer's, whiskered, Brandt's bats and brown long-eared bats. This is not wishful thinking, it happens frequently: seven species flying and feeding in the same airspace. Flying above the harp trap you may see and hear noctules, which are clearly exploiting a different airspace. They are also taking, on average, larger insects than the other species: several studies show that noctules feed predominantly on insects with wingspans greater than 1 centimetre. At the other extreme, the Daubenton's bats fly low, rarely more than 1 metre above the water surface, and typically only 10–20 centimetres above it. They take a substantial proportion of their prey from the water surface itself. However, the diets of these two species are not so different. Both species eat many flies and both also take significant numbers of moths, beetles, lacewings and caddis flies, although Daubenton's bat is more dependent on flies. Of the remaining species, brown long-eared bats stand out as being very distinct, specialising on moths, often taken from surfaces by gleaning. At first glance, all the other species have a diet made up largely of flies, but including varying proportions of moths, beetles, lacewing, caddis flies, bugs and some other groups. There appears to be as much variation between different studies of the same species as there is between species. However, a closer look shows that Natterer's bats frequently have a high proportion of spiders in their diet and recent work shows that they are capable of flying very close to vegetation to catch food. Spiders are also important to whiskered and Brandt's bats, and wing morphology suggests that slow flight and gleaning may be important components of their foraging style. Radio-tracked whiskered bats have been seen to spend a great deal of time foraging close to the ground. Recent work shows that 45 and 55 kHz pipistrelle

Plate 1

(a) Greater horseshoe bat, *Rhinolophus ferrumequinum*, negotiating a narrow fissure in a cave. Bats fly at relatively low speeds and are extremely manoeuvrable.

(b) Greater and lesser horseshoe bats, *Rhinolophus ferrumequinum* and *R. hipposideros* respectively, hibernating in a cave. These are the only British bats to wrap themselves in their wings when torpid.

(c) A cluster of alert greater horseshoe bats, *Rhinolophus ferrumequinum*, in a cave roof. Clustering is used by many bats as part of their thermo-regulation strategy. It may also be important in controlling water loss in hibernation.

Plate 2

(a) Bechstein's bat, *Myotis bechsteinii.*

(b) Natterer's bat, *Myotis nattereri.*

(d) A hibernating whiskered bat, *Myotis mystacinus.*

(c) Daubenton's bat, *Myotis daubentonii.*

(e) Brandt's bat, *Myotis brandtii.*

(f) Torpid Brandt's bats, *Myotis brandtii*, juvenile on the left

Plate 3

(a) Serotine, *Eptesicus serotinus*, one of Britain's larger species.
Confined largely to the south-east of England.

Plate 4

(a) Noctule, *Nyctalus noctula*, calling from a tree stump. A large, fast-flying bat of open habitats.

(b) Leisler's bat, *Nyctalus leisleri*. More common in Ireland than in England, where it replaces the noctule.

(c) Leisler's bat, *Nyctalus leisleri* roosting in a tree crack. Bats make use of a wide range of tree cavities.

Plate 5

(a) 45 pipistrelle, *Pipistrellus pipistrellus*, the most common British bat, together with its sibling species the 55 pipistrelle.

(b) 55 pipistrelle, *Pipistrellus pygmaeus*. Frequently paler than the 45 pipistrelle.

Plate 6

(a) Barbastelle, *Barbastella barbastellus*, in flight. It makes use of a surprising range of habitats and forages over considerable areas.

Plate 7

(a) Brown long-eared bat, *Plecotus auritus.*
Moths make up much of its diet, often taken by gleaning.

(b) Grey long-eared bat, *Plecotus austriacus.* One of our rarest bats,
we know very little about its ecological requirements.

Plate 8

(a) Parti-coloured bat, *Vespertilio murinus.*
A rare vagrant, most recently recorded in 2002.

species have significant differences in diet. Both rely heavily on flies taken in flight, but the 55 kHz pipistrelle takes more insects associated with aquatic habitats.

If we were to examine the bat community of a river valley in the Southwest, with a rich, semi-ancient woodland habitat, we would hope to find more species: perhaps all 16 British natives. However, the additional species, as well as being rare, are all in some way distinctive. Barbastelles are moth specialists, but catch them primarily on the wing, so are perhaps not in direct competition with long-eared bats. Bechstein's bats appear to have a very broad diet. Although flies, moths and beetles are the most important food items, spiders, centipedes, insect larvae, grasshoppers and crickets, cockroaches and earwigs are also eaten in significant numbers. Clearly many of these cannot fly and must be taken by gleaning. The horseshoe bats, in a family by themselves, echolocate with distinctive calls, and the greater horseshoe bat is unique amongst British bats in its habit of feeding from perches, frequently on large beetles and moths. The open-air habitat of the noctule is also occupied by the serotine, but the latter has a greater dependence on beetles than the noctule. Leisler's bat shares the same habitat, and has a similar diet to the noctule. Ireland has a large population of Leisler's bats, but no noctules, in contrast to the situation in England and Wales. Is this circumstantial evidence for competitive exclusion?

So, the picture is one of a community of bats that clearly have overlapping ecological niches, but most species have habitat requirements, foraging styles or food preferences that show some unique features that enable them to coexist with other bats. There is no direct evidence for significant competitive exclusion.

How bats interact with other animals

Bats interact not only with each other, but with other animals too. This section deals with the two most important interactions that take place: bats feed on some animals, and some animals feed on bats. In other parts of the world, other forms of interaction do take place, but these are relatively rare. In Britain it is hard to think of examples of any real significance. Bats do make use of tree holes excavated by woodpeckers, and I know of one old woodpecker hole from which noctules are regularly evicted by nesting starlings. Small birds such as tits and treecreepers regularly use bat boxes (as do wasps), but I do not know if there is competition or conflict.

Are bats in an arms race with their prey?

One aspect of foraging ecology has been left deliberately until now, because it is a complex and fascinating story in its own right. It has been implicit in most of the discussions so far that bats need only to detect their prey to catch them. Of course, it is in the best interests of their prey to avoid detection and capture, and strategies have evolved in insects to help them avoid becoming a bat's next meal. Most bats advertise their presence in a dramatic and unavoidable way: by emitting intense echolocation calls many times per second. Many insects have no way of perceiving sound and have little or no warning of their approaching doom. However, hearing organs have evolved in beetles, bugs, flies, mantids, cockroaches, moths, lacewings, crickets and grasshoppers (Fig. 4.19).

Fig. 4.19 Some of the insects that have evolved 'ears' to detect the echolocation calls of bats.

Sound is a widespread method of communication in insects such as grasshoppers, crickets and cicadas, and the ability to detect bats may have arisen in some cases as a useful spin-off. However, hearing organs in other insects appear to have evolved specifically to detect bats, being most sensitive to sound between 20 and 50 kHz. They also appear to have evolved independently in many of these groups and are very diverse in structure and location: they may be found on the abdomen, thorax, legs, head and wings. Few have been studied in any detail, but most is known about those on moths. Like the hearing organs of other insects studied, moth 'ears' are very simple, usually having just two receptors. The first responds to weak ultrasound, the second to more intense sounds, and each triggers a reflex response in the moth. Low intensity echolocation calls cause the moths to fly away from the sound source. If the second sensor is triggered, in response to high intensity calls, the moth will either suddenly stop flying and drop towards the ground, or begin a series of rapid and unpredictable manoeuvres involving dives, loops and spirals. Both strategies quickly take the moth out of the flight path of a bat. The latter has the additional benefit of appearing to be random and therefore unpredictable, making it more difficult for the bat to intercept its prey. Studies suggest that this behaviour can almost half the chance of capture. There is even evidence to suggest that the rapid repetition of calls in a bat's feeding buzz, independently of the increase in intensity, can alert moths and other insects to their imminent danger.

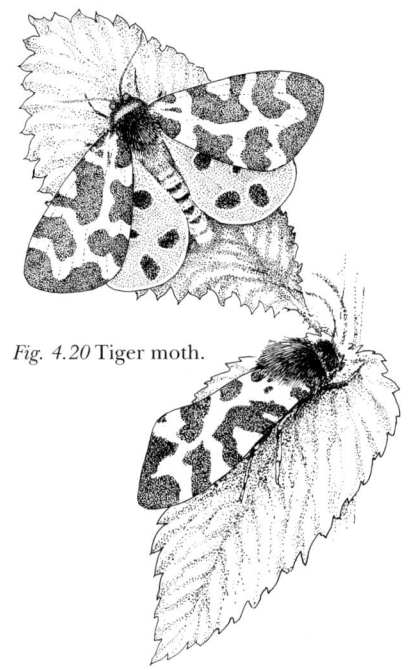

Fig. 4.20 Tiger moth.

Tiger moths (Arctiidae, Fig. 4.20) have evolved another interesting behaviour. In response to the high intensity calls of a nearby bat, some tiger moths emit their own, loud clicks. In some species, these are very similar in structure to bat echolocation calls. This led to the suggestion that they may briefly 'jam' the echolocation calls of the bat, giving the moth an opportunity to evade capture. In support of this idea, it was noted that these clicks were usually emitted during the bat's feeding buzz, when they are likely to be most effective at confounding the bat's echolocation system. Alternatively, a loud sound just as the bat makes its final approach may temporarily startle the bat: even a brief hesitation on the part of the predator may mean the difference between life or death to the prey. There is a third and equally plausible explanation for this behaviour. Tiger moths are frequently unpalatable, and, like many unpleasant or even poisonous animals, they advertise this fact with bright warning colours and patterns. There is no advantage in being poisonous unless potential predators are made aware of the fact. Bright coloration will be a singularly ineffective signal to an echolocating bat in the middle of the night, but audible clicks will be effective. Big brown bats in North America have been shown to learn to associate tiger moth clicks with unpalatability: some young, naïve bats will attempt to catch and eat tiger moths, but will be startled by the loud clicks they emit, and they quickly learn to avoid tiger moths altogether.

The independent origin of hearing organs in so many insects suggests that bats have exerted a strong selective pressure on insect evolution. Hearing, and the escape responses associated with it, could be viewed as a significant step in an arms race between bats and their prey: anywhere between 50 and 90 per cent of moth species in a particular environment may be capable of hearing echolocating bats. Since this can significantly reduce the hunting success of bats, we may expect them to have counterattacked, particularly those species that prey heavily on moths. There are two obvious ways in which bats could bypass the moths' defences. First, they could 'switch off' echolocation and capture their prey by other means. Alternatively, they could use calls that are inaudible to moths, or at least less audible. This can be achieved in several ways: by using calls of low intensity, very short or very long duration, or of frequencies outside the hearing range of moths. All these strategies appear to be used by bats.

Long-eared bats can stop echolocating during the approach, relying on passive hearing to locate their prey in the final attack. Stationary, silent moths are much less likely to be taken than moving individuals. Insects crawling or fluttering against vegetation make sounds that long-eared bats can detect with their large and sensitive ears, and their hearing is particularly sensitive in the appropriate frequency range. Incidentally, the eyes of long-eared bats are characteristically large and I have often wondered whether vision is also used in prey detection. Although vision is important to some New World species, there is, as yet, no evidence that it is important to long-eared bats. The echolocation calls of long-eared bats are unusually quiet and short relative to those of other British species, which makes them less audible to moths and this may be a contributory factor in their ability to predate so heavily on these insects.

Moths are typically most sensitive to sounds between 20 and 50 kHz. The echolocation calls of lesser horseshoe bats, at about 105 kHz, are well above this range and have been shown to be almost undetectable by noctuid moths. At 82 kHz, the calls of greater horseshoe bats are also high, but not high enough to avoid detection. Oddly, greater horseshoe bats rely more heavily on moths as prey than lesser horseshoe bats. Bechstein's and barbastelle bats also prey heavily on moths, but neither have obvious strategies to avoid detection, nor do any other species, most of which take some moths. No British bat has an echolocation call of a low enough frequency to avoid detection. The European free-tailed bat, *Tadarida teniotis*, calls at 11 kHz, and its call may well be inaudible to moths, which make up a large part of its diet.

It can be argued that some of these echolocation calls have evolved for other reasons: high frequency calls may have evolved for the detection of small insects, and low frequency calls for long range detection, as discussed in the previous chapter (p. 45). However, there is no reason why natural selection cannot be operating through several mechanisms. Indeed, natural selection may be all the more powerful when a change in call structure offers more than one advantage to the bat.

Predators of bats

What feeds on bats in these islands? Probably the most important natural predators are owls (Fig. 4.21). It has been estimated that bats represent less than one tenth of one percent of the prey items of British owls, but even this may account for about 10 per cent of annual bat mortality. Most estimates come from owl pellet analysis. Because bats have such delicate skeletons, they are poorly preserved in owl pellets and the numbers taken are probably underestimated. Tawny owls appear to be the most significant predators, and, given the large British population, probably have the largest impact on bat populations. Individual tawny owls have been known to prey quite heavily on bats. At a site near my home, presumed to be the territory of a single tawny owl, bats accounted for almost 5 per cent of the prey items taken over a six year period, an estimated 88 pipistrelles per year, together with a small number of noctules. This is sufficient to have a significant impact on the local population. The evidence suggests that many bats are taken when foraging, but tawny owls have been seen attacking emerging bats and bats returning to the roost at dawn. In Europe, estimates are substantially higher than the British average: from 0 to 4 per cent of prey items taken by tawny owls were bats, with exceptional reports of 9 and 12.5 per cent. One reason for this may be that the owls seem to prefer large bats and most British species are small. Barn owls and long-eared owls also take bats, but apparently fewer than tawny owls, and the more diurnal little and short-eared owls appear to take even fewer.

Fig. 4.21 Tawny owl attacking a long-eared bat.

Diurnal birds of prey probably take the occasional bat, but records are almost exclusively anecdotal. Kestrels and sparrowhawks, given their abundance and habits, are probably the most likely predators and both have been seen to chase bats. It is difficult to estimate their likely impact, but it is probably small.

Roosting and hibernating bats are potentially vulnerable to terrestrial predators, particularly those able to climb, but there are no published studies on this form of predation.

The effect cats have on bat populations is probably vitally important, but since cats are a non-native pest, discussion of them will be left until the conservation chapter (p.139).

Anti-predator strategies

Bats are probably most vulnerable to predation when emerging from, or returning to, the roost. There are several species of raptor around the world that specialise in feeding on bats during the dusk emergence. I have seen a tawny owl attempt to take a pipistrelle as it returned to its nursery roost under the eaves of a building. Many and perhaps most bats taken by cats are caught as they leave or enter the roost. Several behavioural traits seen in bats may have evolved, at least in part, as strategies to reduce the chance of predation near the roost. Small flying insects are usually most abundant at dusk and dawn, and numbers decline, often rapidly, after the sun sets. When bats emerge at dusk, their timing is a compromise between the likelihood of eating and being eaten. Bats that emerge early can take advantage of a more abundant food supply, but diurnal predators may still be hunting and even nocturnal predators may be more effective before it is truly dark. Late emergence may minimise the predation risk, but food will be less abundant. Slow-flying bats and those that glean their prey are more open to predation than fast-flying, aerial foragers. Those that take nocturnally active moths and non-flying prey do not need to emerge early, since they do not depend on the dusk peak of flying insects. The influence of these factors may be seen by looking at the relative emergence times of British bats, in a study conducted specifically to investigate this issue (Table 4.1).

Table 4.1 Average emergence times of British species (From Jones & Rydell, 1994).

Species	Median emergence time (minutes after sunset)
Nyctalus noctula, noctule bat	5
Nyctalus leisleri, Leisler's bat	18
Eptesicus serotinus, serotine bat	20
Rhinolophus ferrumequinum, greater horseshoe bat	25
Rhinolophus hipposideros, lesser horseshoe bat	31
Pipistrellus pipistrellus, common pipistrelle (45 and 55)	32
Myotis mystacinus, whiskered bat	32
Myotis bechsteinii, Bechstein's bat	33
Plecotus auritus, brown long-eared bat	54
Myotis nattereri, Natterer's bat	75
Myotis daubentonii, Daubenton's bat	84

At one end of the scale, the large, fast-flying noctule was the earliest bat to emerge, on average about 5 minutes after sunset. Not far behind were Leisler's bats and serotines (18 and 20 minutes). Noctules and Leisler's bats rely heavily on the dusk peak of small flies. Serotines are more dependent on beetles, but I could find no information on when these are most active. Amongst the latest bats to emerge were the slow-flying long-eared and Natterer's bats (54 and 75 minutes). The first feeds predominantly on moths, the latter has many non-flying prey in its varied diet. Most other species fall between these two groups, but Daubenton's bat was picked out as being unusual in that it emerged latest of all (84 min). Its European relative, the pond bat, *Myotis dasycmene*, also feeds over water and it too was a relatively late riser (64 minutes). One explanation for this is that both are slow flyers relative to other open habitat foragers, which may make them particularly vulnerable to predation. To find statistically significant trends in the data it was necessary to take a European or even a global view, since there was considerable variation between and within species. In our own studies of numerous Daubenton's bat roosts, emergence time was within the range shown by the gleaners and not unusually late.

Emergence from the roost has been shown to be neither uniform nor random, but clustered: bats leave the roost in clearly discernible groups. This clustering behaviour is most easily seen during the emergence of large nursery colonies of more than 100 bats. Several possible reasons have been put forward to explain this behaviour, but the most likely explanation is that it is an anti-predator strategy. By emerging in a cluster of other bats any individual bat is less likely to be the target of a predator than if it emerged alone: this is often referred to as the dilution effect. Although it has not been proven in the case of bats, there is considerable evidence from studies of other animals that there is safety in numbers, and reduced predation pressure is one of the most important benefits of group living. The bats may benefit from more than just the dilution effect, in that a group is more likely to note the presence of a predator than a single individual, and alert the rest of the group. In some circumstances it may also be more difficult for a predator to catch bats in a confusing explosion from the roost. Reduced predation pressure may explain why some bats commute in groups to foraging sites, and may be an additional benefit of group foraging.

It has not been demonstrated in Britain, but studies in other parts of the world have shown that within a species large colonies emerge earlier than small ones. This is probably because individuals benefit from the dilution effect: early emergence is not so risky for an individual that is part of a large group. Remember that individual bats are acting entirely selfishly: they behave in such a way as to minimise their own predation risk and maximise their own feeding opportunities. Another strategy that can reduce predation risk is to switch roost frequently, so that predators are less able to lie in ambush. This strategy has also been demonstrated in subtropical bats, where predation is more easily observed. Once out of the roost, bats often fly to foraging sites under cover, by flying low to the ground or close to hedges and woodland edges: horseshoe bats are very adept at this strategy.

Parasites

Parasites can be considered as another group of animals that feed on bats. The effects of parasites are less visibly dramatic than those of other forms of predation, but they can be a significant burden, reducing growth and survival in the young and fecundity and life span in adults. Bats have both ectoparasites and endoparasites (parasites that live outside and inside the body respectively). Ectoparasites are most common on bats that live in large colonies and those that remain faithful to a single roost, since these traits favour parasite transfer between bats. Young bats frequently have higher parasite burdens than adults, probably because they are less mobile and less efficient at grooming. Bats, like all small mammals, are host to a number of ectoparasites, including fleas, mites and the specialised nycteribid flies. Some do relatively little damage, feeding on exfoliating dead skin, but others penetrate the skin to ingest blood. Many endoparasites, such as nematodes and protozoa, are transmitted by bloodsucking ectoparasites.

5

British Bats, Past and Present

Prehistory and history

Bats evolved around 70–100 million years ago. The number of species increased dramatically about 50 million years ago, at a time when other mammals underwent a similar radiation. There is no evidence to be found of these events in Britain. However, not far away in the oil-shale beds of Messel, near Darmstadt in Germany, some of the world's oldest fossil bats have been found (Fig. 5.1). It is likely that these bats were similar to, if not the same as, the bats present on the part of the globe that was to become Britain. Around 49 million years ago, in the Middle Eocene, the best preserved, oldest and most complete fossil bat community ever found was laid down in Messel. At the time, the area was a region of subtropical lakes and swamps, fed by rivers from a nearby limestone plateau. Fossil bats are usually very rare and incomplete, but at Messel they are by far the most common mammals: several hundred individuals of seven species from three families. These bats may have been overcome by poisonous swamp gases as they hawked for insects over the water. This would explain the huge numbers, the complete, healthy skeletons and the full stomachs of many of the bats. Our good luck does not end there. The oil shales are

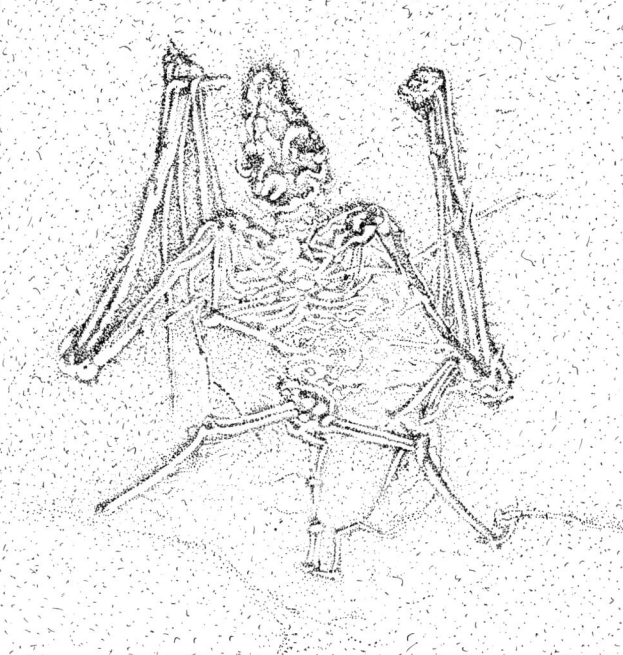

Fig. 5.1
*Palaeochiropteryx
tupaiodon,* one of the
Messel fossil bats.

often very fine-grained and were laid down under anoxic conditions (low levels of oxygen) that prevented decay: ideal conditions for preservation not only of bony skeletons, but of organic components too, including wings and the bats' last meals.

The Messel bats were a highly evolved and diverse community of insectivorous bats. *Archaeonycteris trigonodon* and *A. pollex* were primitive species that still had a claw on the index finger and primitive, unspecialised molars. The term primitive in an evolutionary context means early forms and does not imply inferiority of function. Both bats had forearm lengths around 60 millimetres (a reliable indicator of skeletal size in modern bats), comparable to the modern mouse-eared bat, and were not too dissimilar to modern *Myotis* species in those morphological features that determine flight style and ecology. *Palaeochiropteryx tupaiodon* and *P. spiegeli* were small, similar in size to a lesser horseshoe bat, and had the broad wings and delicate body of many Rhinolophidae, the horseshoe bats. With a low wing loading and low aspect ratio wings they would have been slow, agile and manoeuvrable bats capable of hovering and gleaning in dense vegetation. Because of the excellent preservation of the fossils, it was even possible to analyse the gut contents of some of the specimens. Like many modern horseshoe bats, *P. tupaiodon* fed almost exclusively on Lepidoptera, probably small, weak-flying Microlepidoptera. In addition to Microlepidoptera, remains of caddis fly were found in the guts of *P. spiegeli*. *Hassianycteris magna*, *H. messelensis* and *H. revilliodi* were all large bats, with disproportionately long forearms. *H. magna* was the largest, with a wingspan of almost 50 centimetres. High aspect ratio wings (long and narrow) and high wing loading (a relatively heavy body) made these bats fast and efficient open-air foragers, not unlike modern free-tailed (Molossidae) and sheath-tailed (Emballonuridae) bats. Their diet is less certain, but Lepidoptera and Coleoptera (beetles) were definitely eaten. The variation in wing morphology and hence flight style is almost as broad as that seen in many much larger tropical bat communities today. Interestingly, although *Palaeochiropteryx tupaiodon* has a wing morphology and diet comparable to a horseshoe bat, there is no evidence from X-rays of its skull that it had evolved a similar echolocation system. The inner ear structure is comparable to that of modern FM bats.

The Messel bat fauna is unique: a rare coincidence of the very particular circumstances needed to preserve bats in stone. Fossil bats are very rare in all parts of the world and the British bats are no exception. If the search is confined to Britain, then we have to go a long way forward in time before we find fossils: in fact all the way to the postglacial. Bechstein's bat has been found at Pin Hole Cave in Derbyshire, in deposits dating back to the Neolithic. Bechstein's was the commonest bat at the Neolithic Grimes Graves in Norfolk, and at Ightham Fissures in Kent, a probable late glacial site. In a small cave in the Cresswell Crags area on the Derbyshire–Nottinghamshire border, four species of bat were found: barbastelle, whiskered, Natterer's and brown long-eared. Larger bones from other animals at this site have been dated by radiocarbon techniques to the Mesolithic, 9960 BP. The lesser horseshoe bat has been recorded from four early sites in Derbyshire: the Neolithic Dowel Cave and Pin Hole Cave, the probable Mesolithic Wetton Mill Rock Shelter and Ossom's Eyrie Cave, which is Romano–British (the periods date the finds, not the caves themselves). All these bats are species we would expect or hope to find in British woodland today and fit in with the current view of postglacial

Fig. 5.2 Cresswell Crags, Derbyshire.

Britain as having extensive woodland cover. Daubenton's bat and Leisler's bat have also been found in Dowel Cave, and there are Middle and Late Pleistocene records of serotine from the Nottinghamshire–Derbyshire area and from Somerset. It is particularly significant that Bechstein's bat and the lesser horseshoe bat were apparently so common. These two species are particularly well adapted to, and dependent upon, woodland. After thousands of years of progressive woodland clearance, both species are now rare in Britain.

A look at continental Europe shows a more diverse Pleistocene bat community, as might be expected, but one very similar to that of the present day. Three species of extant horseshoe bat have been found, and perhaps as many as 20 extant vesper bats. Several extinct species of vesper bat have also been described, but all appear to be members of modern European genera. If Pleistocene Europe held no surprises, then it is likely that neither did Pleistocene Britain.

Then comes the documented historical period, when almost certainly there were enormous changes in the British bat fauna, particularly in the last few hundred years, as the increasing human population and technological change put pressure on the land. Unfortunately, very little is known about our bats through this period. Goldsmith, in his *History of the Earth and Animated Nature*, published in 1774, describes an unnamed bat (probably the pipistrelle), the long-eared bat, the 'Horse-shoe' and 'Rhinoceros' bats (lesser and greater or just one of them under two names?). The rest are described as '…. several others, whose varieties are too numerous, and differences too minute for a detail, all are inoffensive, minute, and contemptible: incapable from their size, of injuring mankind, and not sufficiently numerous much to incommode him'. He also says that their 'unsteady wobbling motion, amuse the imagination, and add one figure more to the pleasing group of animated nature'. He concludes that this, and their industry in pursuing insects, more than pays for their occasional sin of entering the larder and stealing the fattest part of the bacon.

Just how sparse even our local knowledge is can be best illustrated with an example, and I will be shamefully parochial and look to my adopted home of Yorkshire. In 1881, Clark and Roebuck published their *Handbook of the Vertebrate Fauna of Yorkshire*. Noctules, long-eared and pipistrelles are described as common or not uncommon. Single records of Leisler's, Natterer's and whiskered bats are described. All records of horseshoe bats are described as unauthenticated and no other species gets even a mention. The position is little better in most other counties.

At the beginning of the last century, Barrett-Hamilton published *A History of British Mammals*, in parts, from 1910, listing 12 British species (Brandt's and the grey long-eared are missing, as well as the 45/55 and Nathusius' pipistrelles). He summarises well the state of play from 1770, when Albin made the first attempts to distinguish between species, recognising three: the common, the double-eared and the (exotic) flying fox. Bingley and Pennant raised the number to six around 1810. The number of species reached a peak with Bell in 1937, by the inclusion of several continental species. In his introduction he says that all but Bechstein's and barbastelle are common somewhere and makes the perceptive point that given the mobility of bats, restrictions in their distribution must be of fundamental importance and worthy of study. This challenge has not yet been taken up in earnest. Barrett-Hamilton's summaries of the distribution of most species vary little from those of the present day and their *relative* abundances are also quite similar. However, there are some interesting departures. In England, Leisler's bat was known only in Yorkshire, Cheshire and the Avon valley. A case is made for the long-eared being more numerous than the pipistrelle in many parts of Britain. The most striking difference is that the lesser horseshoe bat is described as common, with a range extending as a crescent from the Southwest, with Kent and North Yorkshire at its tips.

The present day

Even now, we know rather less than we would like to about the distribution and population sizes of our native species. Not even the number of known species has remained constant in the last ten years, with one becoming extinct and two being 'discovered'. We currently have 16 resident species, representing just two of the world's 17 families of microbats, the Rhinolophidae or horseshoe bats and the Vespertilionidae, the vesper or evening bats. The 69 species of horseshoe bat are distributed throughout the temperate and tropical regions of the Old World. The vesper bats are even more numerous and widespread, with 330–350 species distributed throughout the world. Britain has representatives of six of the 42 genera. These are the only families with more than a small number of temperate species and most other families are confined to the tropics and subtropics.

Western Europe, to the borders with Russia, has at least 34 bat species. There are five horseshoe bats, a solitary molossid and a growing number of vesper bats. Within the vesper bats there is one genus that has never been found in Britain, represented by a single species, Schreiber's long-winged bat, *Miniopterus schreibersii*. All the species found in Britain are present on the European mainland, and many have distributions that extend well beyond Europe to the east. The greater horseshoe bat is found as far east as Japan, as are whiskered, Brandt's, Daubenton's and brown long-eared bats. Most other species are confined largely to Europe. The single molossid is the European free-tailed bat, *Tadarida teniotis*, one member of a genus with an almost world-wide distribution.

The species

The most recent detailed account of British bats is to be found in the *Handbook of British Mammals* (Corbet & Harris, 1991) and those looking for detail and an abundance of source material should look there. Here a brief overview is given, and the species accounts brought up to date. What follows is a separate entry on each species, covering classification, physical description, distribution, population estimate, population trends, conservation status and a basic natural history. Some of the natural history is covered in the previous two chapters and has not been repeated. This section starts with a checklist of those species known to be resident (that is breeding in Britain), together with some vagrant species and species that might be expected in Britain in the future. The North American vagrants are not included, since these are never likely to become established. Climate change is already influencing the distribution of some European animals, including butterflies and birds, and there is no reason to suppose that the distribution of bats will remain static. Distribution maps and population estimates often reflect the local efforts of recorders as much as the real distribution of bats, and even in Britain this is an important factor. The behaviour of individual species is also reflected in its presumed distribution and abundance. For example, large or noisy bats that roost in occupied buildings are more likely to be recorded than those that roost and feed high in the canopies of trees. The population estimates given are usually from a single source, and, although based on the best available methods and data, should be taken as very rough estimates only. Brief physical descriptions of the bats are given, but details are left to the identification key later in the book.

Resident species

Family: Rhinolophidae
        Genus: *Rhinolophus*
                Species: *R. ferrumequinum* Greater horseshoe bat
                         R. hipposideros Lesser horseshoe bat
Family: Vespertilionidae
        Genus: *Myotis*
                Species: *M. bechsteinii* Bechstein's bat
                         M. nattereri Natterer's bat
                         M. daubentonii Daubenton's bat
                         M. mystacinus Whiskered bat
                         M. brandtii Brandt's bat
      Genus: *Eptesicus*
                Species: *E. serotinus* Serotine
      Genus: *Nyctalus*
                Species: *N. noctula* Noctule
                         N. leisleri Leisler's bat
      Genus: *Pipistrellus*
                Species: *P. pipistrellus* 45 kHz pipistrelle
                         P. pygmaeus 55 kHz pipistrelle
                         P. nathusii Nathusius' pipistrelle

Genus: *Plecotus*
    Species: *P. auritus* Brown long-eared bat
            P. austriacus Grey long-eared bat
Genus: *Barbastella*
    Species: *B. barbastellus* Barbastelle

Species extinct to Britain

Family: Vespertilionidae
    Genus: *Myotis*
        Species: *M. myotis* Mouse-eared bat

Vagrant species

Family Vespertilionidae
    Genus *Vespertilio*
        Species:*V. murinus* Parti-coloured bat
    Genus *Eptesicus*
        Species: *E. nilssonii* Northern bat
Several North American and southern European species have been recorded, most being assisted here by humans.

Species to watch for

Species found along the continental coast of the English Channel may turn up in the future:
Family: Vespertilionidae
    Genus: *Myotis*
        Species: *M. dasycneme* Pond bat
            M. emarginatus Notch-eared bat

Resident species

The greater horseshoe bat
Rhinolophus ferrumequinum

Family: Rhinolophidae
The greater horseshoe bat (Fig. 5.3 and Plate 1(a)–(c)) is an extremely widespread species, from Europe and North Africa all the way east to Japan, although its distribution is patchy. It prefers warm temperate climates and is very much at the northern limit of its range in Britain. It may be significant that horseshoe bats often feed through the winter. This is not unknown in other species, but much less common, and may limit the horseshoe bats' northern range.

    The greater horseshoe bat is confined to the south-west of England as far east as Hampshire, and north to parts of southern Wales (Fig. 5.4). Hibernating individuals have been

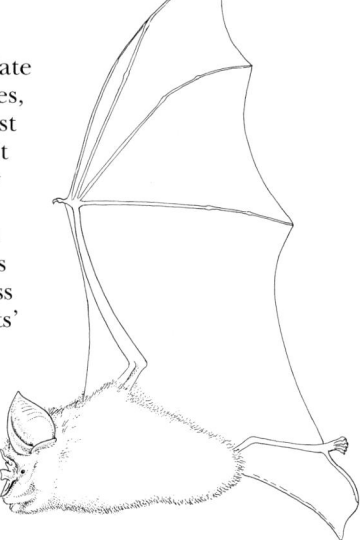

Fig. 5.3 The greater horseshoe bat in flight.

found further east and north. It was reported from Kent and the Isle of Wight in the
nineteenth century. The current population is thought to number no more than
4,000 individuals, 350 of these being in
Wales. It is extremely difficult to estimate
the past population level of any bat, even of
just a few decades ago. Estimates for the
greater horseshoe bat are perhaps amongst
the most reliable, but even so they are very
variable and there is some debate as to how
much the population has declined. It has
been said that in the nineteenth century
there may have been about 300,000 individuals, suggesting a massive loss. However,
this estimate has been criticised as being far
too high. What is not in doubt is that the
decline in the last 100 years has been very
significant, making this bat vulnerable to
extinction.

Fig. 5.4 Distribution of the greater
horseshoe bat.

The greater horseshoe bat is without
doubt one of the most easily recognised
British species for those lucky enough to see it. It is amongst the largest of our
bats, with a typical body mass of 15–30 grams, more commonly at the heavier
end of this range. Its wingspan can approach 400 millimetres. Greater horseshoe bats have a head and body length of about 55 to 70 millimetres and an
average forearm length of about 55 millimetres. Forearm length is a reliable
indicator of skeletal size and an easy and unambiguous measurement to take:
it is therefore widely used both to give a general indication of size, and as an
important element in identification keys. I think of greater horseshoe bats,
together with our other large species, noctules and serotines, as being a com-
fortable handful. Most other British bats are easily lost in a large hand.

Like all horseshoe bats its most obvious feature is its prominent noseleaf
(Plate 1(c)). In addition to serving as a useful feature for identification, the
noseleaf has some very practical functions of importance to the bat rather than
the naturalist. Horseshoe bats emit their echolocation calls through the nose.
The horseshoe-shaped part of the noseleaf, the sella, surrounds the nostrils
and acts as an acoustic lens, helping to focus the emitted sound waves into a
narrow, 30–40° beam in front of the bat. Most of the energy in the calls is therefore directed to that region of space of most concern to the bat – straight in
front of it. This and other functions are discussed in Chapter 3 (p. 39).
Horseshoe bats do not have a tragus (a cartilagenous projection) in the ear.
The ears are large and very mobile (independently mobile) and a roosting bat
will twitch them rapidly when disturbed. The eyes of horseshoe bats are relatively small. The fur is long and dense, typically light brown, darkening and
becoming reddish with age.

In summer, greater horseshoe bats roost predominantly in undisturbed
buildings into which they can fly directly to their roosting positions. Roost sites
with a large range of ambient temperatures are preferred. Females form large
nursery roosts, often of several hundred bats. Males are solitary in summer and

often roost underground. In southern and central Europe this bat will roost and rear young in warm caves in the summer, but this is rare in Britain. When roosting they hang free from the roof, away from possible obstructions to flight (Plate 1(b)).

Females usually begin to breed in their third year and exceptionally can continue breeding for over 20 years, although young are not produced every year. Some females are much more successful mothers than others and it may take eight years for a female to reach its reproductive prime. Young are born in mid-July in most years and birth date is related to spring temperatures. Cool springs lead to late births and high mortality. As with most bats, mortality can be high in the first year or two of life, after which it falls rapidly. Mating occurs in the autumn, as in other bats, when females visit territorial males. Males may occupy the same territory for up to 16 years and may be visited by the same females for several years. Molecular genetics has shown that some males are more successful at siring offspring than others, but the degree of reproductive skew was not great. Males typically fathered 1–3 offspring and up to 40 per cent of the males investigated in a given year fathered pups that year.

Greater horseshoe bats emerge early to forage in deciduous woodland and pastures rich in larger insect species. They feed predominantly on beetles and moths, but take a wide range of other insects. Their diet can change significantly over the summer, as they exploit the seasonal fluctuations in insects such as dung beetles, dor beetles, cockchafers and moths. Foraging habitat needs to be relatively close to good roost sites, with sheltered commuting routes between roost and foraging site. Greater horseshoe bats emerge soon after sunset and feed within 3 kilometres of their roost, at foraging sites they return to most nights. They spend a variable time feeding, depending upon the quality of the habitat and their skill as hunters. Some individuals can rear healthy young on less than three hours foraging per night, but most feed for much of the night, perhaps resting for 1–3 hours in the middle of the night. Most bats divide their time between aerial feeding and flycatching.

Hibernation in the greater horseshoe bat is better understood than in any other species. They favour caves, mines or the cold basements of buildings for hibernation, again choosing sites that enable them to fly right to the roosting position. Torpid bats are very distinctive: horseshoe bats are the only British species to wrap their wings around their bodies when torpid, often hiding everything but the feet and noseleaf (Plate 1(b)). Bats roost alone or in small loose clusters. They prefer temperatures in the range 5–11°C and arouse more frequently at higher temperatures. Arousal is frequent throughout the winter and synchronised to sunset, so that bats can take advantage of warm nights when insects are abundant. In the spring they may emerge to feed at dawn as well as dusk. Monitoring of body mass and faecal production show that greater horseshoe bats do feed through the winter. However many bats arouse without leaving the cave. This may be to mate and drink, or possibly to make physiological adjustments not possible during torpor.

The lesser horseshoe bat
Rhinolophus hipposideros

Family: Rhinolophidae
The lesser horseshoe bat (Fig. 5.5 and Plate 1(b)) has its stronghold in western Europe, but has a patchy distribution into north Africa and the Middle

East. Like the greater horseshoe bat, it is at the
northern limit of its distribution in Britain (Fig.
5.6). It is currently confined to south-west
England as far east as Dorset and north into
the counties bordering Wales, but with occasional records outside this area. It is widely
distributed in Wales. In Ireland it is found in
the western counties. As stated earlier, fossils
have been found as far north as Derbyshire.
In Yorkshire it was apparently quite common
around Nidderdale and Ripon in the late
nineteenth century, but the last confirmed
record was 1894. As recently as 1944 it was still
present in Ryedale in the North York Moors.
There are also early records as far east as
Norfolk and Kent. There are an estimated 17,000 individuals divided between
England (7,000) and Wales (10,000).

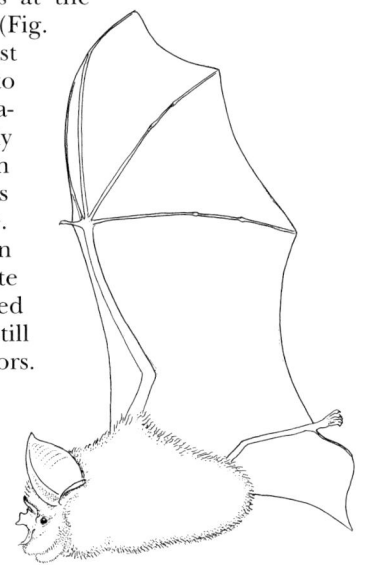

As its name implies, it is a smaller repli-
ca of its larger relative, so much smaller
that the two cannot be confused (Plate
1(b)). Whilst the greater horseshoe bat is
one of our largest species, the lesser is
amongst our smallest. It has a mass of

Fig. 5.5 Lesser horseshoe bat in flight.

around 4–9 grams, a wingspan of 225–250 millimetres, and a forearm length
of 35–43 millimetres. In the hand you are immediately struck by the very delicate nature of this bat: it appears more fragile than vesper bats of a similar size.
It has all the external features of the greater horseshoe bat. Its fur tends to be
longer and does not become red with age.

Fig. 5.6 Distribution of the lesser
horseshoe bat.

Like the greater, the lesser horseshoe bat
appears to have undergone a significant
population decline in the last hundred
years, and a major contraction of its range.
There are no estimates of its previous population, but there are many early, anecdotal records of lost or drastically diminished
colonies. Recent surveys show declines in
the last few decades that may now have
halted.

Roosts in Britain include buildings (from
chimneys and attics down to cellars), caves,
mines and tunnels. If a single building or
building complex offers a wide enough
range of temperatures, then it may be used
all year round. The nursery may be in the
warmer attics, bats will hibernate in the
cooler cellars, and other locations will be
occupied at other times. Like its relatives,
lesser horseshoe bats must be able to fly
directly into the roost and to their roosting

positions. However, they are highly manoeuvrable bats and can negotiate relatively small and tortuous passages. Nursery colonies can number several hundred bats. They emerge early, on average about 30 minutes after sunset. Preferred habitat is sheltered deciduous woodland close to the roost site, with protected flyways to and from the roost. They feed predominantly on flies and small moths, caught in flight close to vegetation or by gleaning.

Bechstein's bat
Myotis bechsteinii
Family: Vespertilionidae
Bechstein's bat (Fig. 5.7 and Plate 2(a)) is rare throughout its range, which is confined to Europe. Even within Europe it is largely absent from the extreme east, much of Spain and Italy, and other Mediterranean regions. In Britain it is one of our rarest bats, confined to an ill-defined area from south-east Wales east to Sussex and Surrey (Fig. 5.8).

Bechstein's bat is a medium-sized bat of 7–13 grams, with a wingspan of 250–300 millimetres and a forearm length of 38–45 millimetres. It has buff to white ventral fur typical of *Myotis* species, with light brown dorsal fur. The ears are large with a long, straight tragus. It has a long, pink muzzle.

Until recently, no more than a handful of bats were recorded each year and no nursery roosts were known. The estimated British population is only 1,500. Serendipity, active searching, concrete bat boxes, and the recent upsurge of interest in studying bats at pre-hibernation swarming sites, have led to an increase in the number of records and an extension of the species' range, but it is too early to revise the population estimate. These new discoveries have

Fig. 5.7 Bechstein's bat in flight and a portrait.

Fig. 5.8 Distribution of Bechstein's bat.

encouraged a number of researchers to work on the species and our knowledge of Bechstein's bat is growing. They are very much woodland bats, requiring mature deciduous woodland with a well-developed three-dimensional structure. They roost in tree holes, generally in the canopy, preferring old trees with dead branches. However, one of Britain's three known colonies is in a house (a rare occurrence), another occupies woodcrete (a mixture of cement and fine wood shavings or course sawdust) bat boxes and the third a veteran oak. The bat box colony, of 50 individuals, moved in just four months after the boxes were erected.

Bechstein's bats emerge about 30 minutes after sunset and feed everywhere from high in the canopy to the ground, and have been observed flycatching from a perch. Their food is primarily moths taken from the air or gleaned from surfaces. There is some evidence that they can locate prey by listening to their movements, in the same way as long-eared bats. Radio-tracked bats from a colony in Dorset had very small, overlapping foraging sites close to the roost. They typically foraged only 300–1,000 metres from the roost in an area of 7–50 hectares: a home range smaller than those of all other British species studied. Nursery colonies are small: up to 50 bats, but usually smaller. They also hibernate in tree holes, although small numbers are found in caves. Recent studies in Germany reported a similar situation of small feeding areas close to the roost. Although many of the females in the colony were closely related, they had largely non-overlapping feeding areas, often distant from each other. Individual bats used the same areas from season to season and even from one year to the next. Young bats fed within the feeding areas of their mothers. Colonies frequently break into smaller units and later fuse, with frequent mixing of the groups, although lactating females tended to stick together.

Natterer's bat
Myotis nattereri

Family: Vespertilionidae
Natterer's bat (Fig. 5.9 and Plate 2(b)) ranges through much of Europe with isolated populations east to the Pacific coast. In Britain it is found everywhere except the north Scottish mainland and the western and northern islands of Scotland. It is also widespread in Ireland (Fig. 5.10). The estimated British population is 100,000, with about 70,000 in England, 17,500 in Scotland and 12,500 in Wales.

Natterer's bat is a medium-sized bat of 6–12 grams with a forearm length of 36–43 millimetres and a wingspan of 250–300 millimetres. It has long, shaggy, light brown dorsal fur and buff-white ventral fur. It has a long, balding pink muzzle. The ears are large and characteristically turned back at the tip, and the tragus is long and pointed.

Natterer's bat is found in a wide range of 'woodland' habitats from open parkland and large gardens to dense woodland, including some coniferous plantations. It roosts in tree holes and buildings, and nursery roosts may have as many as 200 bats. Nursery roosts are not exclusively female: males (adult and immature) may make up 25 per cent or more of the colony. Male-only colonies have been found of up to 30 bats. Like other *Myotis* species it will adopt bat boxes, and even use them as nursery roosts. A study of Natterer's bats using bat boxes in Dorset found small mixed-sex roosts of up to 37 bats, but with an average of only seven. Females outnumbered males by as many as 3:1. Most bats

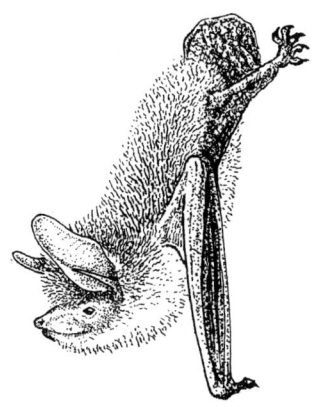

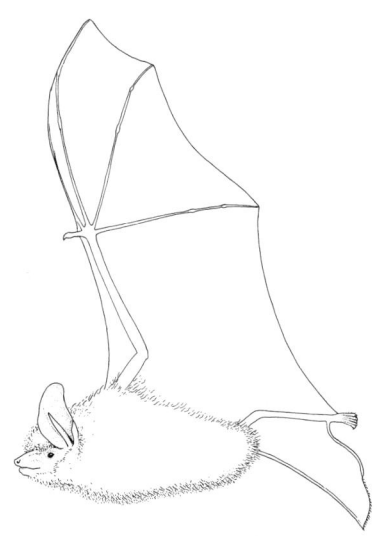

Fig. 5.9 Natterer's bat: a roosting but alert individual, and in flight.

were extremely faithful to a small number of roosts and did not appear to travel far, but a few did move to a site 2.8 kilometres away. Groups of females stuck together, and were often found sharing the same roosts over several years. They have been observed leaving the roost in groups of up to six, but are usually reported as feeding singly.

Natterer's bat is a broad-winged, slow-flying bat that hawks for flying insects. It is also capable of taking prey 2–5 centimetres from vegetation in flight, a behaviour not yet observed in any other species. To do this it uses echolocation calls of unusually broad bandwidth, up to 135 kHz, the FM sweep starting as high as 150 kHz. It probably also gleans prey from vegetation, although it does not appear to land when gleaning, and always emits a terminal buzz, unlike some other gleaners. Its diet reflects its very versatile foraging style: it takes a wide range of insects including flies, moths, beetles, bugs, caddis flies, lacewings and a substantial number of spiders. It also selects relatively large prey (0.5–2 centimetres long) when there is a choice. Even large prey can be eaten in flight when the bat may adopt a characteristic, slow undulating flight, well away from obstacles, perhaps to avoid collisions when echolocation performance is impaired by a full mouth. Prey are often caught in the tail membrane. The stiff, curved bristles along the trailing edge of the tail membrane may have a sensory function and may even stop insects slipping off the edge. Similar bristles are seen on

Fig. 5.10 Distribution of Natterer's bat.

other gleaning *Myotis* species such as Geoffrey's bat, *M. emarginatus,* in Europe and the fringed bat, *M. thysanodes,* in North America.

Natterer's bat hibernates almost exclusively in caves and mines. Together with other species it swarms in late summer and early autumn, usually around cave and mine entrances. Exactly why bats swarm is still uncertain (see Chapter 3, p. 67), but by analogy with work in Canada, mating is probably a major reason. Natterer's bats account for about 80 per cent of all bats captured at swarming sites in the Yorkshire Dales and North York Moors, and they are similarly abundant at swarming sites in Sussex and the Southwest. There is an enormous turnover of hundreds or thousands of bats over just a few weeks of swarming and bats visit from at least 65 kilometres away. Hibernating Natterer's bats are more often than not found tightly squeezed into cracks and crevices, although some hang in the open in undisturbed sites.

Daubenton's bat
Myotis daubentonii

Family: Vespertilionidae

Daubenton's bat (Fig. 5.11 and Plate 2(c)) is found all the way across Europe and Asia, in a narrowing band from Britain, France and the Iberian Peninsula to the Pacific coast and the northern islands of Japan. It is one of the most widely distributed of British bats, being absent only from many of the Western Isles, Orkney and Shetland. It is rare in the far north-west of Scotland, but widely distributed in Ireland (Fig. 5.12). The estimated British population size is 150,000, of which 40,000 are in Scotland and 15,000 in Wales.

Daubenton's bat is medium-sized, with a mass of 7–15 grams, a wingspan of 240–275 millimetres and a forearm length of 33–42 millimetres. It has sleek, uniform brown dorsal fur and pale buff ventral fur. The ears are relatively small, with a relatively short, blunt tragus. Characteristic features are the long

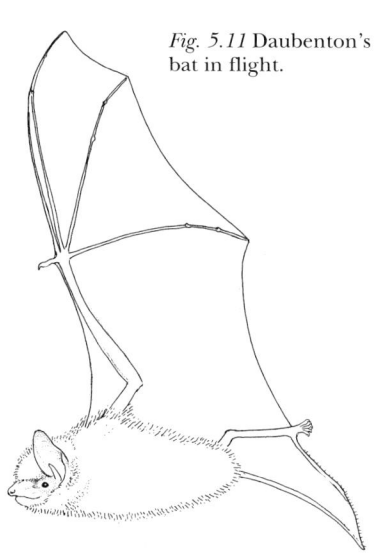

Fig. 5.11 Daubenton's bat in flight.

Fig. 5.12 Distribution of Daubenton's bat.

calcar (the cartilage projection from the foot towards the tail, along the trailing edge of the tail membrane) and the very large feet. *Myotis* bats are undoubtedly the most difficult of British bats to identify, and Daubenton's bat can be confused with whiskered and Brandt's bats by the inexperienced naturalist.

Daubenton's bats rarely roost far from water. Roosts are typically in holes in bankside trees or in the stonework of bridges. Buildings are used, but these are usually within sight of water or at least only a short flight away.

Daubenton's bat is easy to see because it habitually feeds very low over water: ponds, lakes and smooth water surfaces of rivers, streams and canals. It forages almost exclusively in the 1-metre airspace above the water, taking insects out of the air, but also from the surface of the water (Plate 2(c)). The latter are gaffed with the large feet or the tail membrane and quickly transferred to the mouth as the bat continues its flight. Daubenton's bats have been seen to take as much as 40 per cent of their prey from the water surface by this method. Aquatic flies make up most of the diet, but other insects such as Neuroptera and Trichoptera (caddis flies) may be important when abundant. These bats occasionally forage in woodland away from water. Feeding rates can be very high: one study reported an average of a feeding buzz every four seconds for individual bats over water. Studies of Daubenton's bats in the Yorkshire Dales give a quite detailed picture of their natural history. Where available, information from other parts of the country supports this picture. Radio-tracking studies showed that individual bats forage night after night over the same short stretches of river. Most bats have only 1–3 regular feeding sites, which might be as little as 30 metres long and very rarely more than 100 metres. Favoured sites are stretches of smooth water with reasonable tree cover on one or both banks. The bats may be territorial, particularly when insect prey are scarce: foraging sites are often occupied by single bats and rarely by more than three. They are often seen chasing each other, but it has not yet been proven that this is territorial defence. On some occasions 10 or more bats are seen together and chasing behaviour suggests a social function. Most of the bats followed feed for most or all of the night. Males and females emerge from the roost 30–100 minutes after sunset and almost invariably fly direct to one of their chosen foraging sites. Males spend the entire night flying circuits over their feeding areas, only occasionally flying into a nearby tree, maybe five times each night, for five minutes each time. These visits are too short to determine whether or not the bats simply rest in the open or enter a roost cavity in the tree. Either way, this activity is generally described as night roosting. They very rarely return to the day roost before the end of the night's foraging, which typically lasts for six hours, but varies between 3.5 and 8.5 hours. Before their young are born, females rarely use night roosts, but lactating females not only returned to the day roost, but also occasionally made short visits to night roosts. Visits to the day roost were presumably to suckle young. One or two such visits are made each night, often with remarkably regular timing from night to night, each visit lasting an average of 45 minutes.

One roost can be used for weeks on end, but all bats are capable of switching roost frequently, staying in a particular site for as little as one night at a time. Even lactating females used day roosts away from the main nursery roost for one or more consecutive nights. On at least some of these occasions they appear to be transferring their young to these alternative roosts since bats have been seen flying with young several times. In one to two weeks of continual tracking an individual bat will use up to three roosts. Studies in Europe report even more frequent roost switching. No work has been done to determine why they move, but it may be to roost closer to currently rich foraging sites, to use roosts with more favourable microclimates, or perhaps to avoid parasites.

Although they typically forage within 3 kilometres of the roost, Daubenton's bats are quite capable of flying 15 kilometres or more along a river during a night's activity, at flight speeds exceeding 20 kilometres per hour. This behaviour appears to be more common in males, perhaps because they are seeking females. This behaviour may be related to one of the more surprising findings: not only do males and females use different roosts, but their foraging sites are geographically separate. At least in the Yorkshire Dales, the higher reaches of a dale are occupied only by male bats. Habitat segregation may be quite widespread amongst bats. It has been observed in Daubenton's bats in Wales and Switzerland and in the little brown bat in Canada. A recent study showed a dramatic reduction in the proportion of reproductive females (of 11 species) with increasing altitude in the Black Hills of Dakota. I suspect it will be seen more often in Britain if we look for it. Although we do not have the large elevational gradients seen in many other countries, the climatic effects of elevation are greater at these latitudes. This will be compounded by the fact that so many of our bats are clearly at the northern edge of their distributions where small climatic effects can have major consequences. This segregation is of considerable importance to our understanding of the biology of this and perhaps other species, and to their conservation.

It is often assumed that male bats roost in small groups, but male Daubenton's bat roosts can exceed 60 individuals in these upland dales. The nursery roosts lower down the dale are not exclusively female: about 25 per cent of the bats may be males, many of them mature adults. Nursery roosts can hold several hundred bats, but may have fewer than 20. A colony of bats uses several nursery roosts and there may be frequent fragmentation and movement between roosts.

Behaviour patterns become increasingly unpredictable during August and September as the young gain independence and the nursery colonies break up. Daubenton's bat is a common swarming species, so mating probably occurs at traditional pre-hibernation swarming sites, but mating in summer roosts late in the season cannot be ruled out. Mating is also known to occur throughout hibernation. Daubenton's bats, like other *Myotis* species, usually hibernate in underground sites such as caves, mines and suitable tunnels, and may emerge and fly during winter. They will hibernate on the open walls of caves as well as in crevices.

Whiskered bat
Myotis mystacinus

Family: Vespertilionidae
The whiskered bat (Fig. 5.13 and Plate 2(d)) is found in north temperate latitudes all the way from Europe to Japan. It is widespread in England and Wales, but in Scotland it is found only in the south (Fig. 5.14). Its distribution must be a little uncertain, since it was not separated from Brandt's bat until about 30 years ago. New records are extending its range north and west. It is widespread in Ireland. The British population is believed to be about 40,000, with

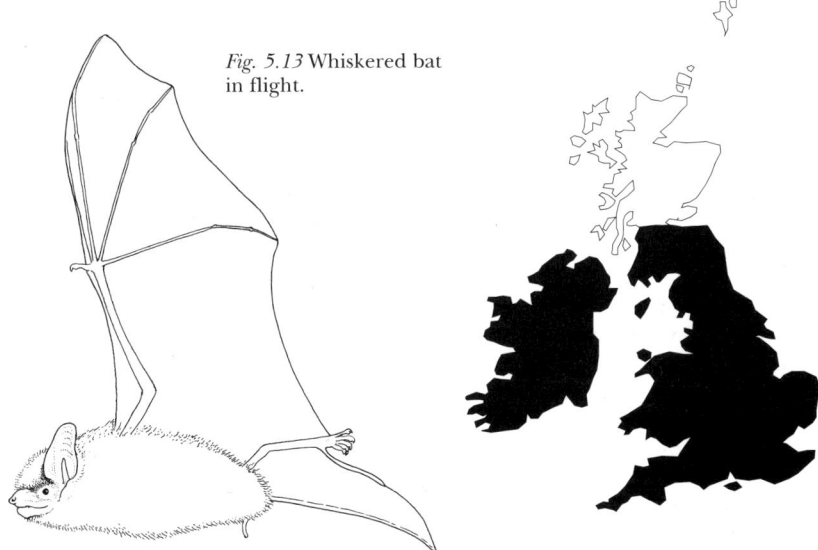

Fig. 5.13 Whiskered bat in flight.

Fig. 5.14 Distribution of the whiskered bat.

8,000 of these in Wales and as few as 1,500 in the south of Scotland.

The whiskered bat is the smallest of our *Myotis* species, along with Brandt's bat, at 4–8 grams, with a wingspan of 210–240 millimetres and a forearm length of 30–37 millimetres. It has dark brown, sometimes almost black, dorsal fur and buff/white ventral fur, a dark face, and medium-sized ears with a long, pointed tragus. It can be very difficult to distinguish it from Brandt's bat and it is easily confused with Daubenton's bat by the inexperienced naturalist.

The whiskered bat is found wherever there are trees, but it favours woodland and riparian habitats. It forages along tree lines, hedgerows, in woodland clearings, gardens and along rivers. Recent radio-tracking studies suggest that it may glean much of its prey, and its wing morphology and diet support this idea. Although flies are the major component of its diet, it eats many spiders. It also takes moths, beetles, caddis flies, lacewings and bugs in significant numbers. It roosts in trees and a wide range of building types. Nursery roosts, almost exclusively made up of females, can number over 100 bats. Whiskered bats hibernate singly in caves, on the open wall or in crevices (Plate 2(d)). Like other *Myotis* species, it swarms in and around caves in the autumn, perhaps a little earlier than the more abundant Natterer's bat.

Brandt's bat
Myotis brandtii

Family: Vespertilionidae

Brandt's bat (Fig. 5.15 and Plate 2(e)) was only separated from the whiskered bat in 1971. Its distribution is therefore uncertain. In Britain it appears to have a similar distribution to the whiskered bat, but it may not extend far into Scotland (Fig. 5.16). Unusually, it appears to be more common in the north of England than the south, but this pattern may change as more bats are recorded. It is absent from Ireland. The British population is thought to be about

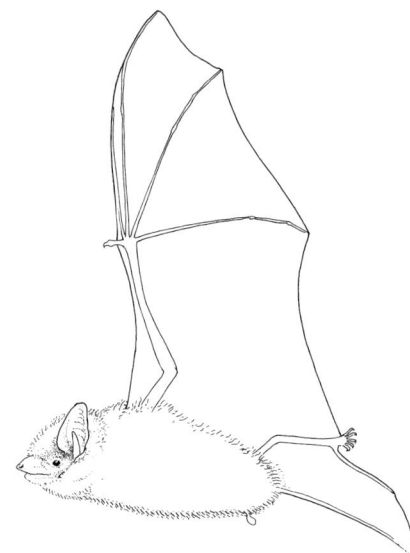

Fig. 5.15 Brandt's bat in flight and at rest.

30,000, 7,000 of these in Wales and perhaps just a few hundred in southern Scotland.

Brandt's and whiskered bats can be difficult to distinguish. Brandt's bat is usually a lighter brown, but positive identification must be based on several subtle features (p. 195).

Summer and winter roosting habits and foraging ecology and behaviour are apparently similar to those of the whiskered bat, but more research is needed on both species.

Serotine
Eptesicus serotinus

Fig. 5.16 Distribution of Brandt's bat.

Family: Vespertilionidae

The serotine (Fig. 5.17 and Plate 3) is widely but patchily distributed across temperate Europe and Asia. In Britain it is found largely in the south-east of England, with only occasional records north of a line joining the Wash and the Bristol Channel (Fig. 5.18). There have been a few recent records in south Lancashire, but there is no evidence to suggest that this expansion (if real) reflects an increase in population. There is in fact some evidence to suggest this species is declining, and new records may simply reflect increased recording effort. It is not found in Ireland. The current British population is thought to be about 15,000.

The serotine is a large, well-built bat of 15–35 grams, with a wingspan of 320–380 millimetres and a forearm length of 48–55 millimetres. The dorsal fur is long and dark brown, grading to lighter brown on the ventral surface. The

Fig. 5.17 Serotine in flight and a portrait.   *Fig. 5.18* Distribution of the serotine.

hairs may have pale tips. The ears are relatively short with a short, blunt-tipped tragus. The ears and muzzle are dark brown to black.

This is a bat primarily of lowland parkland, pasture and woodland edge. In Sussex, bats from one roost fed over nearby chalk scrubland and made use of recently mown grass fields when chafers were emerging. It flies relatively slowly on its broad wings, taking prey from above the tree canopy to down at ground level, but typically flies about 3–4 metres from the ground. It feeds mainly on beetles, often taken during sweeps close to the ground, but also takes moths, a range of flies, and other insects in smaller numbers. Some of the insects taken are diurnal and the serotine has been seen to glean these from the ground. One study reports serotines feeding in short flights from a perch on a building or tree. Small groups have been seen feeding around mercury-vapour street lamps, in company with noctules and pipistrelles. Home ranges for the species can be large: from less than 1 square kilometre to almost 50 square kilometres, and the home range for a colony of up to 20 bats can approach 80 square kilometres. Individual bats commute on average 8 kilometres between the roost and two to three foraging sites on a single night, but as many as ten sites can be used and total commuting distances can exceed 40 kilometres. Commuting routes usually follow hedges, woodland and pasture.

Before humans destroyed most of its natural roosts, the serotine presumably roosted in tree cavities. It now relies almost exclusively on buildings, preferring small cavities and crevices, with high access points, in older buildings. It is therefore inconspicuous despite its large size. Very occasionally it is found in trees or bat and bird boxes. Nursery colonies are small, typically up to 30 females and males roost singly or in small groups. Reproductive females are usually faithful to a single roost for long periods, but females without young can change roost frequently. Emerging bats typically fly direct to foraging sites, but early in the season they may return to the roost within 40 minutes. As the season progresses, the time out of the roost increases and activity is recorded throughout the night. In common with most other species, returning bats frequently spend many minutes circling the roost entrance, occasionally landing, before entering the roost.

The serotine hibernates in buildings, but also uses caves, mines and cellars. It appears to be a relatively sedentary species, travelling only short distances to hibernation sites if it moves at all.

Noctule
Nyctalus noctula

Family: Vespertilionidae
The noctule (Fig. 5.19 and Plate 4(a)) is widespread throughout Europe, with a patchy distribution around the Mediterranean and east to the Pacific Ocean. In Britain it is found throughout England and Wales and in the extreme south-west of Scotland (Fig. 5.20). It is absent from Ireland. There are thought to be around 50,000 bats in Britain, 5,000 of these being in Wales, with perhaps a few hundred in Scotland.

The noctule is an easily recognised large, sleek bat. With a body mass of 15–49 grams, a wingspan of 330–450 millimetres and a forearm length of 47–58 millimetres, a large noctule is larger than any other British bat. It has short golden or ginger fur and a well-groomed appearance. The ears are short with a short mushroom-shaped tragus. The ears and muzzle are dark. The fur extends onto the wings along the body and arms.

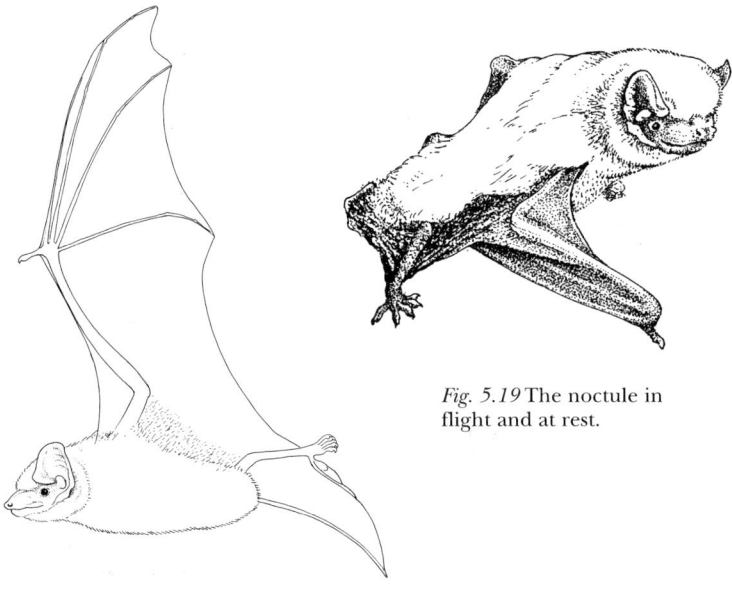

Fig. 5.19 The noctule in flight and at rest.

Fig. 5.20 Distribution of the noctule.

The noctule is found in a range of habitats, and, as a large, fast flier is capable of travelling long distances between roosts and foraging sites. It is one of the earliest bats to emerge and forage and can be seen well before dark. It will often hunt for only 60–90 minutes before returning to the roost, but may feed again around dawn. It feeds in the open, often over trees, but it shows a strong affinity to water. It has a rapid, efficient flight on long narrow wings, characterised by frequent stoops and glides. It feeds mostly on flies, beetles and moths, taking predominantly large insects, but some surprisingly small ones.

Noctules roost almost exclusively in tree holes in Britain, although they are found in buildings and other man-made structures. Tree-hole roosts can be as little as 1 metre from the ground. They occasionally use bat boxes. Buildings are used more extensively in continental Europe. Nursery colonies rarely exceed 20 females in Britain, but can be much larger in continental Europe. Males roost singly or in small groups. In late summer and autumn individual males establish territorial mating roosts from where they make distinctive calls, presumably to attract females, and harems containing up to 18 females have been found. In some parts of its European range these females are not present in summer, but migrate into the region to mate and hibernate. Trees are also the most common hibernation sites, although buildings may again be used. Occasional noctules have been caught entering caves during the autumn. Large numbers of bats may cluster within a tree hole, enabling them to tolerate very low ambient temperatures. In eastern Europe as many as 1,000 bats have been found in the same roost. This species is migratory in eastern Europe and has been known to travel over 2,000 kilometres. In western Europe and Britain it is either non-migratory or migrates over much shorter distances.

Leisler's bat
Nyctalus leisleri

Family: Vespertilionidae

Leisler's bat (Fig. 5.21 and Plate 4(b)–(c)) is confined largely to southern and central Europe, with a patchy distribution in the west. It is rare in Britain, being found in small numbers as far north as Yorkshire, with a cluster of records in south-west Scotland. Interestingly, although apparently absent from Wales, is it very common in Ireland where it replaces its larger relative the noctule (Fig. 5.22). The estimated British population is 10,000.

Leisler's bat looks like a small version of the noctule: its body mass is 11–20 grams, wingspan 260–340 millimetres and forearm length 38–47 millimetres. Apart from size, the major difference is that the short dorsal fur is bicoloured, with dark roots and red-brown tips.

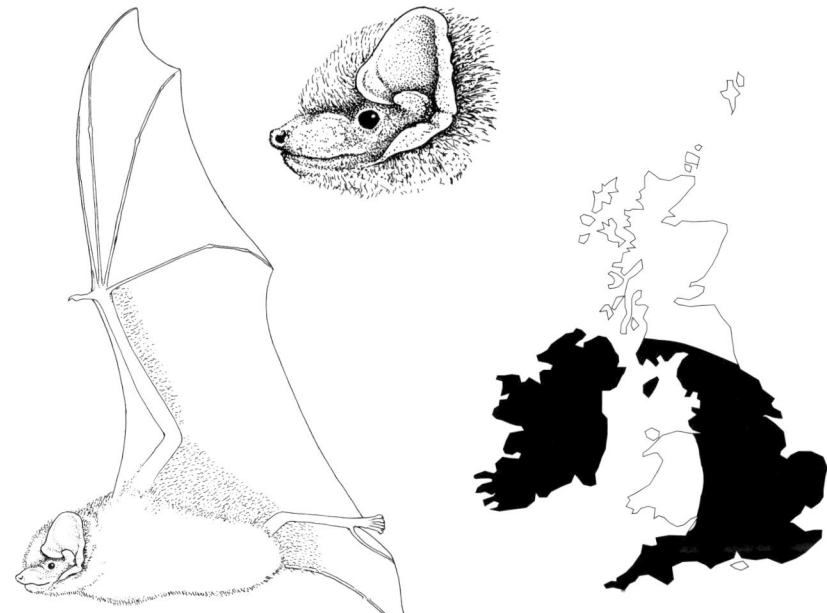

Fig. 5.21 Leisler's bat in flight and a portrait.   *Fig. 5.22* Distribution of Leisler's bat.

Leisler's bat is not so dependent as the noctule on tree roosts, using a wide range of buildings too. A colony will use several alternative day roosts in buildings and trees. In Ireland, nursery colonies can number over 500 females. Average home range area can approach 18 square kilometres and foraging flights can take it up to 13 kilometres from the roost. It feeds primarily on flies, but will take smaller numbers of other insects, in particular beetles and moths, both of which can comprise up to 20 per cent of the diet at certain times of the year. It often forages for up to two hours and then returns to the roost, making second or even third foraging trips later in the night, particularly when feeding conditions are favourable. Radio-tracked bats in Kent foraged preferentially along woodland margins, even when these were along major roads. In other areas a range of foraging sites were selected, including woodland, pasture and riparian habitats. In Ireland it prefers pasture, also makes substantial use of riparian and woodland habitats, and even forages over beaches and sand dunes.

Male Leisler's bats call to attract females in late summer, either from a perch or in flight. This behaviour was best described in a study carried out in Greece. In summer only males were caught in the northern, mountain study site, females coming from roosts even further north at the end of the summer. Males typically call in the early morning, before sunrise, and use the same perch night after night. A male will sing for 2–10 minutes at a time, repeating a 30-millisecond narrowband FM sweep (18–10 kHz) once every second for long periods.

Little is known about its hibernation biology, but Leisler's bat probably prefers tree holes like the noctule. Like the noctule, it is also migratory over part of its range.

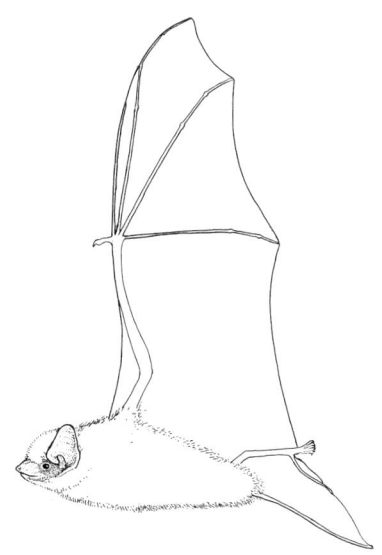

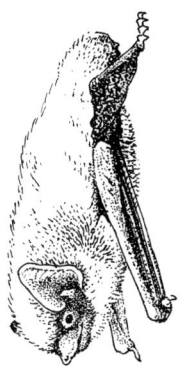

Fig. 5.23 45 kHz pipistrelle in flight and at rest.

45 kHz pipistrelle
Pipistrellus pipistrellus

Family: Vespertilionidae

Describing many aspects of the natural history of our most common species is now fraught with difficulty. Virtually all studies carried out up until the last few years assumed that there was a single, common species of pipistrelle in Britain. Few people would now dispute the evidence that divides it in two. Much of what was known about the 'combined' species can probably be applied with reasonable accuracy to both new species, but already significant differences have been discovered. I will describe the natural history of the old 'common' pipistrelle, *P. pipistrellus* (Fig. 5.23 and Plate 5(a)), making distinctions where new information permits it. This, I hope, will result in a reasonably accurate account of the new *P. pipistrellus*. The new *P. pygmaeus* (Fig. 5.25 and Plate 5(b)) will have a brief but separate account, highlighting known differences.

The first problem is one of names. The most widely used labels have been 45 kHz and 55 kHz pipistrelles for *P. pipistrellus* and *P. pygmaeus* respectively. The first and crucial clues towards their eventual separation were the very different frequencies of the pseudo-CF terminal phase of their echolocation calls. This is still the most reliable method of separating the two species and these names are the most descriptive. Alternative names are in use and all have their champions and detractors. I think the names '45s' and '55s' are snappy, descriptive, unambiguous and have historic significance, and I use them here in preference to any of the alternatives. Even the specific name *pygmaeus* has not been universally accepted: recent publications from Europe refer to the 55 as *P. pygmaeus/mediterraneus*.

The combined species are widespread throughout Europe and the Near East. They undoubtedly coexist over much of their range, but 45 pipistrelles appear less common in the north. In contrast, they appear to be the most abundant in the south, and the range of the 55 pipistrelle may not extend as far south as that of the 45.

Both 45 and 55 pipistrelles are found in all parts of Britain and Ireland (Fig. 5.24), but again 55 pipistrelles may be more abundant in the north, and 45 pipistrelles more abundant in the south. 45s were known to be present in Orkney in the early 1990s. The estimated, combined population is two million and the two species appear to be roughly equally abundant. They remain the two most common British species by a substantial margin.

The 45 pipistrelle is a small bat with a body mass of 3.5–8.5 grams, a wingspan of 180–250 millimetres and a forearm length of 28–35 millimetres. The dorsal fur is almost black at the base, turning to brown at the tip. The ears are short and black and the tragus short, curved and blunt. The muzzle appears longer than in the 55 pipistrelle. The muzzle is also black and the bare, black skin extends back around the eyes to the base of the ear giving the bat a distinctive black mask. The penis fur is grey.

Fig. 5.24 Distribution of the 45/55 pipistrelles.

Their abundance must be due in part to their adaptability: pipistrelles are found everywhere from upland areas to big cities. They forage wherever there is sufficient vegetation to support an adequate insect population, from woodland and rivers to suburban gardens and city parks. They emerge early, often in daylight and, like most species, fly directly to their foraging sites, which are typically within 2 kilometres of the roost, but may be as far as 5 kilometres away. They hunt over water, along tree and hedge lines, along woodland edges, in gardens and even around solitary trees. Flight is fast and agile, with frequent, rapid changes of direction. The diet is primarily small flies. When insects are scarce, pipistrelles may become territorial and defend food patches, using social calls and chasing to warn off other bats.

Nursery roosts are almost invariably found in buildings and a wide range of structures is used from ancient churches to new houses. Almost any small cavity within a building will be used and they occasionally form clusters in an open roof space. Nursery roosts can number more than 200 females, but are often much smaller. A colony may use several roost sites, sometimes moving between them within a single season. Large colonies may fragment and occupy several roosts at the same time. Pipistrelle roosts can approach and even exceed 1,000 bats, but most large colonies have proven to be of 55 pipistrelles. Males roost solitarily or in small groups.

In the late summer and autumn solitary males occupy mating roosts within small territories that are defended against other males. They have a distinctive songflight that they use to attract females, which spend the following day in the males' roosts. The songflight call (p. 66, Fig. 3.22) is similar to the social calls recorded at other times (see Chapter 7, p. 168), but detailed analysis reveals some differences. However, these may be due to the fact that both sexes produce social calls and only males songflight. Whilst calling, the male flies an elliptical path around his roost, always approaching or landing briefly at the roost as if to draw its attention to passing females. The more time a male spends in songflight display, the more females he can attract: sometimes more than ten females at a time, but typically one to three. It is not unusual for a male to spend 40 per cent of his time in songflight activity after feeding for an hour or so. When songflighting and mating, males lose weight, but this is rapidly regained when mating ends towards the end of September.

Although pipistrelles may once have relied more heavily on trees, tree nursery roosts are now uncommon. Solitary males use tree holes, and regularly use bat boxes as mating roosts. Groups of 50 or more females use bat boxes in the spring before moving on to nursery roosts in buildings. Few winter roosts in Britain are known and all are occupied by solitary or small groups of bats. They are often very exposed sites in external cracks and cervices in buildings. In continental Europe, large roosts of hundreds or thousands of bats are not uncommon, in old buildings, caves, mines, chalk pit and cellars. In Britain they are rarely found in caves, but significant numbers of pipistrelles have been recorded at autumn swarming sites in the north-west corner of the Yorkshire Dales. Hibernating pipistrelles appear relatively insensitive both to cold and to large diurnal temperature fluctuations. Most populations appear to be non-migrating, moving less than 20 kilometres from summer roosts to hibernation sites. Occasional movements up to 770 kilometres have been recorded in Europe.

55 kHz pipistrelle
Pipistrellus pygmaeus

Family: Vespertilionidae

Although there is considerable overlap, the 55 pipistrelle (Fig. 5.25 and Plate 5(b)) may be more abundant in northern Britain than the 45 pipistrelle.

It is a small bat, overlapping with the 45 pipistrelle in all measurements, but on average it is very slightly smaller. The main distinguishing features are its paler, single colour fur and its pink face. Other features that often aid identification are a somewhat flatter snout, larger swellings around the nostrils and yellow fur on the penis.

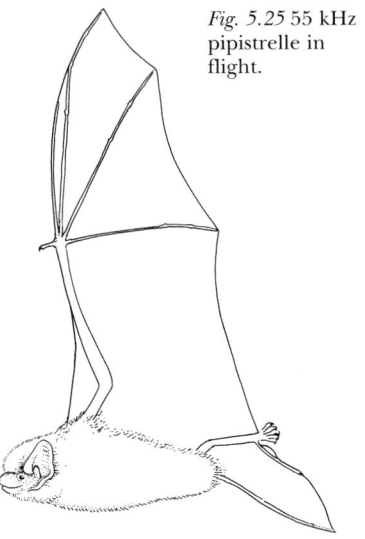

Fig. 5.25 55 kHz pipistrelle in flight.

Although it has similar foraging habits to the 45 pipistrelle, the 55 pipistrelle is more closely associated with riparian habitat and this is reflected in its diet, which has a greater proportion of aquatic flies. The 55 pipistrelle usually forms larger nursery roosts than the 45 pipistrelle. In Scotland roosts of 500–700 bats are not uncommon and some roosts exceed 1,000. Although the 45 pipistrelle shows a high degree of roost fidelity, the 55 pipistrelle is even less nomadic, frequently using the same roost for the entire season and often year after year.

Nathusius' pipistrelle
Pipistrellus nathusii

Family: Vespertilionidae

Nathusius' pipistrelle (Fig. 5.26) is widely but patchily distributed in Europe, with the greatest numbers in the east. It was thought to be a vagrant to Britain, with only occasional records up to 1985, most of them on or near the south and east coasts. There have been many more observations since then, most frequently in May and September, suggesting migratory individuals. One bat found on Jersey had been ringed in Germany. This led to the suggestion that some of these bats hibernated in Britain after setting up mating territories along their migration route. More recently, two very young individuals were found in Cambridgeshire and 'singing' males were subsequently recorded in south-west England. Nursery roosts were then discovered in Northern Ireland and Lincolnshire, another roost in the south-west of England, and individuals are turning up more frequently inland. Because it is so easily confused with the more common 45 and 55 pipistrelles it may have been present, but overlooked, for some considerable time, although this is unlikely given the professional and amateur interest in bats over the last two decades. Alternatively Nathusius' pipistrelle may be extending its range into Britain, an expansion perhaps related to climate change. The distribution of as yet undiscovered colonies may help to resolve this, although the recent expansion theory does not fit well with the current roost distribution pattern. That said, nursery colonies have only recently been found in the Netherlands and the species has reappeared in Spain and Portugal after a long gap, lending support to this theory. Whatever the reason for its recent appearance, there is no doubt that Nathusius' pipistrelle should now be included amongst our resident species. Although migratory over its continental range, in the mild climate of Britain it may be more sedentary.

Nathusius' pipistrelle is bigger than the more common species, but the size ranges do overlap: its body mass is 6–15 grams, wingspan 220–250 and forearm length 32–37 millimetres. The winter coat of long, shaggy dorsal fur with 'frosted' white tips and significantly lighter ventral fur can be distinctive. In summer the coat is less distinctive. The most reliable identification features are given in the key on p. 193.

Nathusius' pipistrelle is a woodland bat, preferring lowland woods and parks. Its foraging style is similar to that of other pipistrelles, but it appears less agile. The diet is primarily flies. Nursery roosts in Europe are almost always in tree holes, but it

Fig. 5.26 Nathusius' pipistrelle in flight and a portrait.

will readily use bat boxes and crevices in wooden fire towers or shooting towers. It rarely uses buildings in continental Europe, but the known British nursery roosts are in buildings. It hibernates in hollow trees and also in crevices in cliffs, walls and caves. Like other pipistrelles the males appear to use a distinctive call to attract females to their autumn mating roosts. However, they usually call from a perch close to the roost, rather than in flight.

It is a migratory species, typically flying south-west from summer roosts to hibernacula, often over 1,000 kilometres, with a maximum recorded migration of 1,600 kilometres.

Brown long-eared bat
Plecotus auritus

Family: Vespertilionidae

The brown long-eared bat (Fig. 5.27 and Plate 7(a)) is found throughout Europe and is patchily distributed east to Japan. In Britain it is found everywhere except the more remote Western Isles and Orkney (Fig. 5.28). There is one record from Shetland. It is widely distributed in Ireland. After the two common pipistrelle species it is probably our next most abundant bat, with an estimated British population of about 200,000. Almost 30,000 of these may be in Scotland and 20,000 in Wales.

Long-eared bats cannot be confused with any other British species. These medium-sized bats (mass 6–12 grams, wingspan 230–285 millimetres, forearm length 34–42 millimetres) have enormous ears, 29–41 millimetres in length. The inner margins of the ears meet in the middle of the forehead and the tragus is large and prominent. At rest the ears are often curved back in the shape of ram's horns, and can be folded right back and tucked under the wings during sleep, leaving the erect tragi looking like much smaller ears. The eyes are proportionally much larger than those of other British bats. The dorsal fur of the brown long-eared bat is long and light brown to yellow, grading to cream on the ventral side. The muzzle is long and usually bare and pink.

It prefers open woodland and parkland and is also found in towns and cities with large gardens and trees.

Fig. 5.27 Brown long-eared bat in flight and at rest.

It has short, broad (low aspect ratio) wings for slow, manoeuvrable flight, and habitually flies close to vegetation hawking for insects and gleaning them from surfaces. It frequently hovers and uses its large ears to listen for the sounds made by moving prey. The large eyes suggest that sight may be important in prey detection, as it is in some other gleaning species. Studies show that moths are always an important component of this bat's diet. Since many moths can hear the ultrasound of an approaching bat, the ability to locate them without echolocation is clearly important. Long-eared bats leave the roost later than most species. Slow-flying, gleaning species may be more likely to be taken by diurnal or crepuscular predators. They usually stay out of the roost all night and return about an hour before sunrise. Night roosts are used throughout the night, usually for brief periods, but sometimes for an hour or

Fig. 5.28 Distribution of the brown long-eared bat.

more. Prey are often taken to night roosts or feeding perches to be eaten, and discarded wings and other bits of food often give away the location of these sites. Like many species they disperse from the roost along regularly used flyways: typically hedges and tree lines. They use these flyways to visit a variable number of feeding sites, often returning to the same sites on consecutive nights. Males use more sites on average than females and travel further to these sites. However, all bats stay relatively close to the roost, most frequently feeding within 0.5 kilometres, often up to 1.5 kilometres away, but never further away than 3 kilometres. This is surprisingly similar to the pattern observed in Daubenton's bat, despite its very different foraging strategy. This is discussed in greater detail in Chapter 4 (p. 90).

Nursery roosts are found in trees (presumably their traditional sites), in the roof voids of large and often old buildings, and in bat boxes. Buildings with roosts have significantly warmer roof spaces than other buildings nearby and also have significantly more woodland within a 500-metre radius, highlighting the importance of this habitat. Nearby woodland also enables long-eared bats to emerge earlier, perhaps due to reduced predation risk. Long-eared bats also prefer buildings with large and more complex roof voids, with timber lining (sarking, a common feature in old Scottish buildings). The large roof voids probably give a wide range of microclimates and the sarking provides better insulation against external temperature fluctuations. Nurseries are usually composed of 10–50 bats (up to 100) and the sexes are not segregated. Long-eared bats show a high degree of roost fidelity. Long-term studies of building and bat box roosts show that less than 1 per cent of ringed, recaptured bats leave the roosts at which they were first captured.

Long-eared bats make up a small but significant proportion of bats at underground swarming sites. As with other species, swarming males outnumber females. They hibernate in tree holes, buildings and in caves and mines. When using underground sites, they are most frequently found near entrances where temperatures are lower. They are non-migratory.

Grey long-eared bat
Plecotus austriacus

Family: Vespertilionidae

Globally, the grey long-eared bat (Fig. 5.29 and Plate 7(b)) has a similar distribution to the brown long-eared bat, but is considerably less common. In Britain it is confined to a coastal strip of Devon, Dorset, Hampshire and Sussex, along with the Channel Islands (Fig. 5.30). Several colonies in Dorset and Devon have declined to extinction in the last 30 years. The total British population may be only 1,000 bats. This species is right on the edge of its northern distribution here, and has probably never had more than a tenuous hold on Britain. It is absent from Ireland.

It is very similar in size and appearance to the brown long-eared bat, and fur colour is an extremely poor guide to identity, although it is more likely to be grey than otherwise. It has a body mass of 7–14 grams, wingspan of 255–300 millimetres and a forearm length of 37–45 millimetres. The most reliable differences are its wider tragus and shorter thumb. It was only recognised in Britain in 1963 and its natural history appears very similar to that of the brown long-eared bat: the minor differences occasionally commented upon require more study.

Fig. 5.29 Grey long-eared bat in flight.

Fig. 5.30 Distribution of the grey long-eared bat.

The barbastelle
Barbastella barbastellus

Family: Vespertilionidae
The barbastelle (Fig. 5.31 and Plate 6) is confined to Europe and is sparsely distributed throughout its range. In Britain there may be as many as 5,000 bats and it may be present throughout England and Wales, but it is so rarely seen that even its distribution is uncertain. It is absent from Ireland. All recent records fall south of a line between the Humber and the Mersey, including a few in South and central Wales (Fig. 5.32). It is almost certainly more common in the south than the north. The most northerly records are from Ryedale in Yorkshire, the last in 1956, where a significant colony appears to have existed. A single bat was caught in York in 1968. Until recently, no nursery colonies were known, but in the last few years they have been found in Norfolk, Sussex, Devon and Somerset, sparking off new research.

The barbastelle is a very distinctive bat. It is medium sized: mass 6–13 grams, wingspan 260–290 millimetres and forearm length 36–44 millimetres. It has dark brown/black fur 'frosted' with yellow or cream tips. The face is short and flat and framed by large, almost square ears that are joined together at the forehead. The tragus is large and triangular.

In Europe it prefers wooded countryside, forming small nursery roosts (10–20 females, rarely up to 100) in buildings and occasionally in tree holes and cracks. Males roost separately in small groups. Radio-tracking studies reveal a strong preference for richly structured forests. In Britain the majority of roosts are in trees, frequently dead and storm-damaged. In a Sussex study, on emergence the barbastelles often foraged under the canopy until it was dark before flying rapidly to foraging sites over water and trees and along

Fig. 5.31 Barbastelle in flight and a portrait.   *Fig. 5.32* Distribution of the barbastelle.

woodland edges and avenues. All foraging sites were within 250 metres of a watercourse. A study in Somerset found them foraging in more open country, with scrubby trees and gorse a favoured foraging habitat. In the same study, roosts were located in cracks in trees and under bark, as little as 1 metre from the ground. Home ranges appear to be surprisingly large for a bat with a wing morphology that suggests slow flight. It feeds primarily on small moths that are generally caught in flight.

The barbastelle is largely non-migratory and hibernates in buildings, caves, mines, cellars and tree holes. Some eastern European hibernacula contain clusters of over 1,000 bats. It is a very cold-tolerant species.

Species extinct in Britain

Mouse-eared bat
Myotis myotis
Family: Vespertilionidae
The mouse-eared bat (Fig. 5.33) is distributed throughout Europe except Scandinavia and the North and Baltic Sea coasts. A small colony was discovered in Sussex in 1969, but this was probably a tenuous hold on the country. The loss of the last individual of this colony in 1990 was certainly a cause for regret, but was perhaps not so surprising. What was surprising was the appearance in 2001 of another dying individual in Bognor Regis on the West Sussex coast.

The mouse-eared bat is a very large bat, with a mass of 20–45 grams, a wingspan of 365–450 millimetres and a forearm length of 57–68 millimetres. Its size alone distinguishes it from all British bats. Its dorsal fur is light to medium-brown, and, like most *Myotis* species, it has buff/white ventral fur. Juveniles, as in many species, have greyer fur. The face is relatively bare and pink or light brown and it has the long muzzle of *Myotis* bats. The ears are very long and there is a long, straight, pointed tragus.

For a description of its natural history, we must look to European accounts. It is a bat of thinly wooded countryside. Females form large nursery colonies of 100 or more, with a few adult males present. The females visit solitary males in the autumn to mate. Nursery colonies are in buildings or warmer caves and cooler caves are used for hibernation. It feeds primarily on larger moths and beetles.

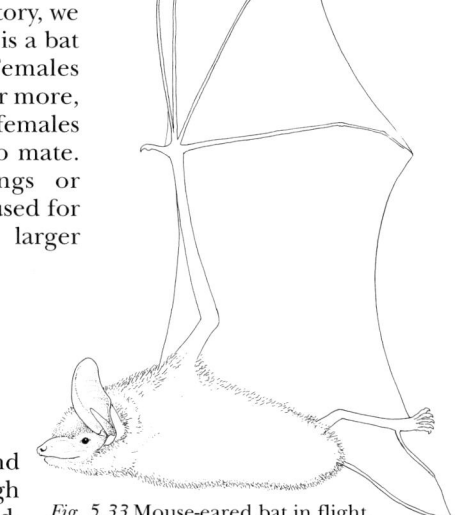

Vagrant species

Parti-coloured bat
Vespertilio murinus
Family: Vespertilionidae
The parti-coloured bat (Fig. 5.34 and Plate 8) is sparsely distributed through central and eastern Europe, extend-

Fig. 5.33 Mouse-eared bat in flight.

ing to 60°N. The occasional vagrants to
Britain (the most recent in 2001), where
it has been found from Shetland to
Plymouth, are probably lost migrants.

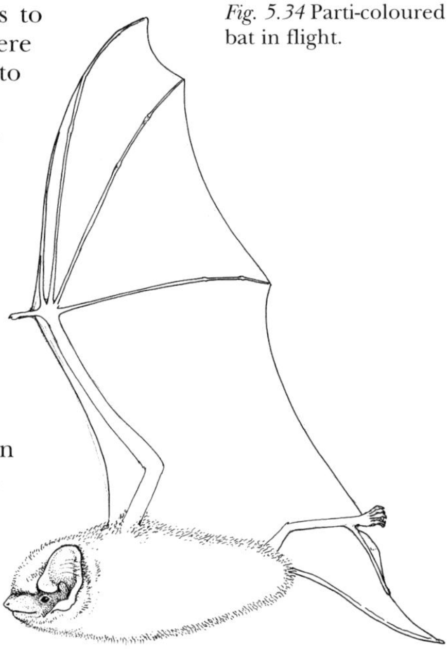

Fig. 5.34 Parti-coloured
bat in flight.

The parti-coloured bat is medium
sized with a body mass of 11–24
grams, wingspan of 260–330 millimetres and forearm length of
39–49 millimetres. Its dense, dorsal fur is characteristically black/
brown at the base with silver/white
tips. The fur on the underside is
white or grey. The ears are short and
broad and the tragus short and mushroom-shaped.

Summer roosts are typically in
crevices, in walls, windows and attics,
including tower blocks. It has been
suggested that it may originally
have been a cliff dweller. Nursery
colonies are relatively small (<50
females), but males can congregate
in roosts of 250 individuals in summer. This is a fast-flying, open space
forager. In Sweden and Poland, small flies make up as much as 80 per cent of
the diet (by volume), but moths and beetles can be important. Moths become
more significant when the bats forage around street lamps, a common behaviour in northern Europe. Hibernating bats prefer
crevices in caves and cellars. It is a migratory
species over at least part of its range, travelling up
to 900 kilometres to hibernation sites predominantly south-west of its summer roosts. Long
and complex mating calls have been recorded.

Northern bat
Eptesicus nilssonii

Family: Vespertilionidae
The northern bat (Fig. 5.35) is the only
European bat to be found near and indeed
above the Arctic Circle. Its range extends
south to cover much of northern and eastern Europe, and east to Japan. It has been
recorded only once in Britain, in a
hibernation site in Surrey.

The northern bat is similar to its
relative the serotine, but much
smaller, with a body mass of 8–18

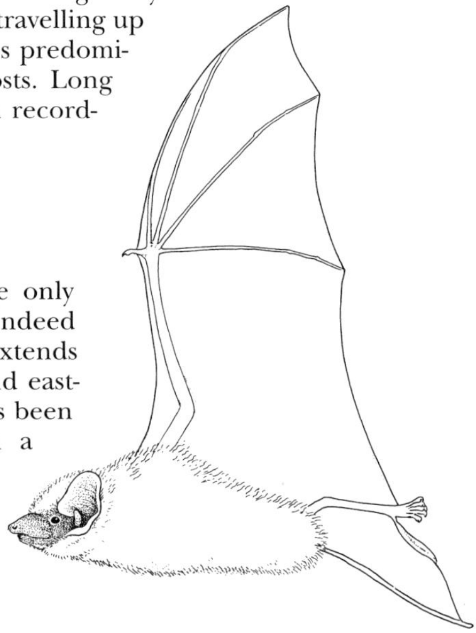

Fig. 5.35 Northern bat in flight.

grams, a wingspan of 240–280 millimetres and a forearm length of 37–46 millimetres. The long, dark brown dorsal fur is distinctively gold-tipped.

The northern bat is found in a wide range of habitats from forest and woodland to farmland. It forages in open spaces and along woodland edges, with a fast and agile flight. It feeds primarily on small flies. Nursery roosts are typically in cracks and crevices around buildings. Solitary bats are also found in tree holes. It is primarily a non-migratory species, hibernating in caves, mines and tunnels, hanging in the open or using crevices.

6

Conservation

Despite the fact that bats make up about one third of our native, terrestrial mammal species, they received relatively little attention until quite recently. An unfortunate consequence has been a largely unnoticed decline in their populations. We can barely begin to estimate how much their numbers have fallen, since current population estimates are unreliable and past numbers unknown. By the time people began to take an interest in bats and promote their conservation, there was already substantial if anecdotal evidence to show that our bat fauna was already under enormous pressure. We can only guess at the number of bats that could have been sustained by the rich and abundant habitats of the past, but we can be sure that it was many more than in our present degraded countryside. Anecdotal though much of the evidence may be, it is nevertheless strong. There are numerous records of once large bat colonies that no longer exist or are sad remnants of their former selves. In a few cases, the declines are well documented. The geographical range of several species has clearly contracted in the last 100 years. Bats observed frequently by local naturalists a hundred years ago are no longer found in some areas, or are seen rarely. I can only dream of seeing two species at my study sites in Yorkshire that were described as quite common in the area not long before I was born. The picture is similar across Europe and North America, where similar environmental pressures have been at work, and across the rest of the world, where the reasons for declines are often more varied. From even a simple understanding of bat biology it is known that those features of the environment essential to bats' survival are being fragmented, degraded and destroyed. The assault on our biodiversity in general is very well documented and it can be confidently inferred that bats must be part of this decline.

Threats to bats

What are the threats to British bats? Some are common to much of our wildlife, others are almost unique to bats, due either to their unique natural history or their public image. The main threats are habitat fragmentation, degradation and loss, roost disturbance and destruction, and deliberate persecution.

Habitat use: commuting and foraging

Most of our bats are best adapted for feeding in a landscape that is a mosaic of woodland, still and slow-moving water and open areas of grass and heath. The woodland should have a complex three-dimensional structure, with areas of dense understorey, clearings and edges. The water should be clean with a rich bankside vegetation of trees, shrubs and smaller plants. The open spaces should have a herbivore community to help maintain this diverse landscape. All should be composed of native plant species with an abundant and diverse invertebrate community. This landscape has the complexity in habitat and

Fig. 6.1 Ideal bat habitat.

food essential to provide each species with commuting corridors, foraging
beats and a sustained food supply.

   Whilst this type of habitat can still be found in Britain, it is increasingly confined to small, isolated fragments, and sometimes only preserved in nature
reserves. Semi-ancient, deciduous woodland is now much less common than it
was a century ago. It has been replaced by farmland, buildings and coniferous
plantations. Even where woodland exists the structure has often been lost:
even-aged stands under which overgrazing has stopped regeneration, removed
the understorey and destroyed the diversity of the ground flora. Coniferous
plantations are made up largely of non-native species, which do not have the
rich insect communities associated with native trees. They are also frequently
monocultural, simplifying structure and making them prone to attack by pests
and diseases: pests and diseases that may then be treated chemically. Old and
dead trees are often removed to limit the spread of disease, to avoid damage
and injury due to falls, or out of some misplaced sense of tidiness. Open areas
are predominantly either intensively arable or heavily grazed and fertilised
grassland. Biologically rich, unimproved grassland has been ploughed up and
sown with less diverse and often non-native grasses. Fences have taken the
place of hedgerows and fields are large to accommodate large machines and
efficient harvesting practices. Shelter for feeding and commuting is harder to
find. All these changes have reduced the diversity and abundance of feeding
sites and of food essential to the bats. Insect abundance and diversity depends
upon the presence of the plants and microhabitats that sustain them, and both
are in short supply. The insects that remained have been further reduced by

chemical warfare: herbicides destroy the weed plants that feed the insects that feed the bats. Insecticides either kill the bats directly or accumulate in their bodies from contaminated food. This kills them slowly, reduces their fecundity and may be passed on to their suckling offspring.

Other changes in farming practice can also have an impact. Fewer cattle are turned out into fields, reducing the dung essential to the beetles on which several bat species feed. The dung that is available may be contaminated by drugs fed to the cattle to control internal parasites. These drugs poison the insects that try to make use of the dung. Artificial fertilisers reduce the diversity of plants in the ground through which they pass, again reducing insect diversity. When the fertilisers get to the streams and rivers they lead to eutrophication (increased nutrient cycling and plant growth) which reduces insect diversity and biomass. Although the underlying causes depend on the nature of the contamination, urban and industrial sewage can have similar effects and reduce the foraging activity of bats. It has been argued that sewage can increase insect abundance and benefit some bats. However, an abundance of a limited number of insect types, perhaps only available in certain months and at certain times of night, is no substitute for a diverse and relatively undisturbed invertebrate fauna. Many of our rivers and streams rush through this landscape between straightened banks covered in low, species-poor vegetation. Fast-flowing and disturbed, murky water often has fewer insects, and the turbulence and noise reduce the effectiveness of echolocation. Insect-rich wetlands have been drained for centuries and converted into largely sterile arable land. Ironically, at least for some of our more common bats, many urban and suburban areas provide better foraging than the countryside.

Summer roosting

Bats spend a very large part of their lives in the roost and good roosting sites are as important as good feeding sites. Not only have we removed most of our woodland cover, but little of what is left provides natural roosts. Suitable cavities are most common in old trees: windblown, cracked, lightning-struck and rotten veterans. To many people such trees are ugly, untidy and a danger to humans. Commercial foresters may see them as a source of disease, so they are often cut down and removed. Although there are far fewer suitable tree roosts than in the past, humans have provided a huge diversity of alternatives in built structures. There is little doubt that some species, such as pipistrelles, serotines, horseshoe and long-eared bats have adapted well to such sites. However, these roosts do need to be largely undisturbed and close to good feeding sites. Roosts in buildings may be torn down, renovated, or disturbed unintentionally. The treatment of timber with chemicals to kill woodworm and other wood-boring pests has been a major concern. Until recently the most frequently used chemicals used were organochlorine pesticides and gamma-HCH (for example, Dieldrin, PCP and Lindane). These are highly toxic, highly persistent, readily absorbed and have killed many bats. Some have now been withdrawn because of their danger to humans, but others are still in use. Until recently bats were often regarded as pests and either excluded or destroyed without a second thought.

Many species have taken little or no advantage of buildings as roosts, such as Bechstein's and barbastelle bats, and it may be no coincidence that these are now amongst our rarest species. Old, disused buildings, such as castles and monastic ruins, are important to many species, which use both rooms and wall cavities: unsympathetic restoration can exclude or even entomb bats. Crevices in the stonework of bridges are particularly important to Daubenton's bats, but are also used by other species. Several surveys have shown that up to 30 per cent of bridges in a given area may be used by bats. In recent years pressure injection of cement into cracks has been used in the repair of many bridges: a technique that can entomb whole colonies.

Hibernation

Hibernating bats are particularly vulnerable for several reasons. Good hibernation sites appear to be rare and many therefore attract large numbers of bats. The loss of single sites can have drastic consequences. Hibernating bats may take 30 minutes or more to arouse from torpor and escape from danger. Disturbed bats can use up vital energy reserves that cannot be replaced in the winter. Many bats use structures described in the previous section for hibernation, such as buildings, castles and bridges, all needing sympathetic management. Those species that hibernate underground probably have a greater number and variety of hibernation sites now than for most of the last 10,000 years. Many have adapted well to mines, disused railway and canal tunnels, ice houses and similar structures and some of these are now important and protected hibernacula. However, these sites are often vulnerable to damage and disturbance through vandalism, dumping or innocent curiosity. Many have been sealed to protect people and livestock. Tourism and caving for fun must have been responsible for many losses in the past, and even biologists entering caves to study bats can have a serious effect.

Introduced predators: cats

Regrettably, the family cat is probably a major conservation issue. Given that 7.5 million of these predators are turned out into our gardens and countryside every day and night, and that another one million feral, and therefore hungrier, cats roam the country, it is inconceivable that their impact can be anything other than substantial and detrimental. By way of comparison, the most important natural predators are probably owls. Add together all owls of all species in Britain and you get a number less than 100,000.

A recent survey by the Mammal Society was based on a sample of 1,000 cats, countrywide, over the summer of 1997. The results included only 'what the cat brought in' and ignored what it ate or left outside. Leaving aside this substantial hidden kill, it still concluded that cats killed about 230,000 bats each year. That is equivalent to more than the entire population of any species other than the two most common pipistrelles. If these 1,000 cats are typical, and there is no reason to believe they are not, cats kill many more bats than all natural predators combined. They are one of the biggest causes of bat mortality in Britain, perhaps the biggest.

Other studies support the Mammal Society's findings: cat-inflicted injuries account for the vast majority of dead and injured bats brought to British vets. A recent study of the effects of wildlife gardening on wildlife showed that you are significantly less likely to see small mammals in your garden if cats are present. Considerable efforts are being made to protect bats and their habitats, but little is said about the problem of cats. It has been argued that cats are only an urban problem, but this is not true. Cats may be less abundant in the country, but their densities are still high, much higher than natural predators in most areas. Furthermore, as our countryside has been degraded, village, suburban and even some urban environments have become increasingly important to bats. The only common measure taken to control cat predation is the attachment of bells to cat collars: the Mammal Society study and several studies in the United States and Australia have shown convincingly that this is ineffective. In North America, Australia and New Zealand, where cats have been a significant factor in the decline and extinction of some native mammals, cat predation on wildlife is a seriously debated issue. Cats are acknowledged as a major conservation problem in many parts of the world, directly responsible for the loss of many species, so why is the problem largely ignored here in Britain? Why do most conservation organisations tiptoe around the issue? Is the subject too controversial in this nation of cat lovers? Is it too much to ask that cats, like dogs, be given less freedom to roam?

Climate change

Finally, a factor that may be more fundamental, and certainly more difficult to control: climate change. The evidence for an increase in global temperature due to greenhouse gases is now widely accepted amongst both scientists and politicians. The climate is known to undergo major fluctuations for entirely natural reasons and it is possible that the current trends are entirely natural, a point grasped by sceptics and some with a vested interest in ignoring the problem. However, the theories on climate change are now very sophisticated and predictions fit the observed changes in the climate so well, that few doubt that humans are now having a significant impact on our planet. Given the possible consequences, it seems sensible to be cautious, assume that climate change is largely caused by human activity, and act to halt and reverse the changes. What would climate change mean to British bats? As with many other organisms, it is hard to be sure. An increase in temperature will make winters less harsh, summer roosts warmer and perhaps insects more abundant. But climate change is not simply global warming. Most models suggest that in Britain it will be accompanied by an increase in rainfall and stormy weather. If this is the case, foraging may be more difficult and roosts in old trees less abundant or less secure. This is a simplistic view of a complex issue, but it would be better not to see this vast natural experiment run to its conclusion to see whether the predictions are reliable.

Conservation priorities

Fortunately, this huge list of threats has been acknowledged and the problems are being addressed. Later this chapter considers the many advances that have been, and are being, made to counteract this grim catalogue of destruction. Progress has been considerable, but many threats are still very potent and none have been eliminated: there is no evidence to suggest that conservationists can relax. On the contrary, the Government continues to sanction the destruction of woods and wetlands, and even to drive major roads through nature reserves. The Common Agricultural Policy leads to the investment of huge sums of money in habitat destruction for often questionable or even non-existent benefits. After the tribulations of the 2001 foot-and-mouth epidemic there were promises of radical reform and new, environmentally sensitive strategies for the countryside. There is little sign yet of these promises being fulfilled. Subtle but cumulative damage occurs every day across the country: habitat degradation usually occurs by a process of piecemeal destruction and neglect. We need a strong conservation plan and the resources to implement it. Recent governments, including the present one, have been largely ineffective. There is plenty of talk and political posturing, but little action and little money. Immediate political expediency appears to have a greater influence than real concern for the future, and the debate is too frequently portrayed as conservation versus economic well-being, when both are surely possible. Both practical conservation and conservation research are severely underfunded.

It is difficult to assess the relative importance of the many threats to bats. Without a doubt, priorities change with species and location. However, it is important that conservation works on a broad front. It is not enough to protect a roost, or even all roosts, if foraging habitat is progressively degraded and destroyed. Conservation strategies should make as much as possible of our countryside and towns suitable for wildlife *and* the needs of humans. Conservation should be by consensus and sometimes compromise if it is to be widely accepted and sustained. 'Fencing off' small areas of land for special protection is of enormous importance in special circumstances, but must be just one approach. Whether key sites for bats are protected, or wider areas of the countryside conserved and restored, the best habitats for bats are amongst the richest for other organisms too, so bat conservation rarely needs to be distinct from other conservation work.

In an ideal world, all species would be equally important and equal parts of our conservation effort would be devoted to all. In reality, compromises must be made. Criteria for conservation must be established, a list of priorities drawn up and our limited resources concentrated on what are considered to be the most important problems, those that can realistically be tackled, and that will give maximum returns on the investment. Clearly this is a difficult equation to balance. Do we put our limited resources into a handful of rare species, possibly doomed to extinction (here if not abroad), or bats as a whole? Do we concentrate on practical conservation or raising public awareness of the problems? Do we give priority to practical conservation where we know what to do or to conservation research? The answer, of course, is that we need to do all these things. But there is still the problem of slicing the cake: how many slices and how big should they all be? The next sections consider many steps that are being taken towards bat conservation.

UK and European legislation

Legislation has both followed and facilitated the blossoming of bat conservation in recent decades. The first legal protection given to bats was through *The Conservation of Wild Creatures and Wild Plants Act 1975*, which gave limited protection to the all but extinct greater mouse-eared bat and the greater horseshoe bat. To fulfil the requirements of the *Convention on the Conservation of European Wildlife and Natural Habitats* (Bern, 1982), *The Wildlife and Countryside Act 1981* was drawn up. This act gives protection to all species. Under the Act it is illegal to intentionally kill, injure or take a bat, to possess a bat (alive or dead), or to disturb a bat when roosting. It is also illegal to sell or offer to sell any bat, alive or dead. Where appropriate, these activities may be carried out under licence. The law does permit the care and subsequent release of an injured bat or the killing of a bat injured beyond reasonable chance of recovery. Because of their particular dependence on roost sites it is also an offence to damage, destroy or obstruct access to the roost site irrespective of whether bats are present or not. This protection applies to all roosts in buildings with the exception of bats in the living area of a house, when bats may be removed without injury. If it is considered essential to evict or disturb bats the relevant statutory nature conservation agency must be notified. They must then be given reasonable time to advise on whether the proposed action should be carried out, and if so when and by what methods. The Act also called for the designation of Sites of Special Scientific Interest (SSSIs or Areas of Special Scientific Interest, ASSIs, in Northern Ireland) to give protection to important areas of land.

More recent European legislation has also been of importance to bat conservation, such as the *Directive on the Conservation of Natural Habitats and of Wild Flora and Fauna* (1992). All species receive full protection under Annex IV. Five species (greater and lesser horseshoe, Bechstein's, barbastelle and mouse-eared bats) are included in Annex II that calls for the creation of Special Areas of Conservation (SACs) to protect them (known collectively as Natura 2000). The most recent has been the Convention on the Conservation of Migratory Species of Wild Animals. As far as bats are concerned, its main function has been to encourage member states to set up 'Agreements', such as the *Agreement on the Conservation of Bats in Europe* (1994). This made various statements about the protection of key habitats, co-ordinating research and increasing public awareness, but involved no changes in UK legislation.

The Rio Earth Summit in 1992 led to *The Convention on Biological Diversity*, committing signatory nations to develop strategies for the protection and sustainable use of biodiversity. The UK Government published *Biodiversity: The UK Action Plan* (DoE, 1994). This has resulted in the production of action plans for six species: pipistrelle, greater and lesser horseshoe, Bechstein's, barbastelle and greater mouse-eared bats. The species chosen are unusual in that they include our most common species, the pipistrelle, together with the mouse-eared bat, a species that probably never had more than a tenuous hold on the UK and that is now extinct. No provision is made for many species of uncertain status and even more uncertain population trends.

The most recent piece of legislation is *The Countryside and Rights of Way Act* (2000), known as the CROW Act. It has been heralded as a major change in government policy towards conservation and wildlife: from passive documentation and advising to practical conservation and enforcement. On the enforcement side, it has given teeth to the statutory nature conservation organisations. One of the major problems of *The Wildlife and Countryside Act 1981* was that activities against bats, such as killing or roost destruction, had to be shown to be 'intentional' for a successful prosecution and this was virtually impossible to prove in court. It is now necessary to show that an act was 'reckless' and this is a concept well known in legal circles and more easily proven. Penalties have also been increased and police powers extended. The law is not all it could have been, but it should prove to be significantly better at deterring acts against bats and their roosts. It is harder to see how the many other clauses relating to wildlife in the act will affect bat conservation, but some could be important. Prior to CROW, biodiversity was a direct concern only of the Department of Environment, Transport and the Regions (DETR). Other departments had no responsibilities for environmental issues and often showed little or no regard for the concerns even of DETR (and now DEFRA (Department of Environment, Food and Rural Affairs)). Now all departments must 'have regard to the purpose of conserving and enhancing biodiversity in accordance with the Rio Biodiversity Convention'. It remains to be seen how well this will work, since to succeed it needs commitment on the part of ministers and civil servants. The Act also requires the promotion by government of action to conserve and enhance threatened species and habitats. As it is phrased, this is essentially legal backing for the Biodiversity Action Plan process.

Bat conservation organisations

Until the 1980s very few people were involved in the study or conservation of bats in the UK. Although interest was already growing, the appearance of *The Wildlife and Countryside Act 1981* almost certainly acted as a catalyst. Projects launched by the Fauna and Flora Preservation Society (now Fauna and Flora International) provided a focus for those interested in bats and the Act gave the then Nature Conservancy Council limited legal teeth and limited resources for conservation work. Around the country local bat groups were formed, which came together under the umbrella of Bat Groups of Britain. Out of this grew the Bat Conservation Trust. Representatives of the bat groups meet with those from government agencies, NGOs (non-governmental organisations) and others on a regular basis to discuss the legal and practical aspects of bat conservation. The Bat Conservation Trust hosts an annual conference with lectures and workshops on all aspects of bat biology and conservation, at home and abroad. This strong partnership of groups and individuals has made enormous strides in bat conservation over the last 20 years, influencing conservation in Britain and the rest of the world. Several other organisations have played significant roles in bat conservation in Britain, including the Mammal Society, the National Trust, the Vincent Wildlife Trust, The People's Trust for Endangered Species, the World Wide Fund for Nature, local and county Wildlife Trusts and many others.

Survey and monitoring

Survey and monitoring are cornerstones of good conservation. Survey work grades into scientific research and there is no clear boundary, but a separation is made here for the sake of convenience and research will be discussed later. To conserve a group of animals effectively it is necessary to know what species there are, how many there are, where they live and how their populations change. It is also necessary to know their essential requirements in terms of food and shelter and perhaps understand some of the subtlety of their ecology and behaviour. At this stage I will consider only *what* is being done, and the relevance of the information. The next chapter will look at *how* we can go about obtaining the information.

How many bat species do we have in the country? Are they all found elsewhere in the world, or are some of them unique to Britain? If not unique, might we hold a significant part of the world's population, or a unique assemblage of species? These are largely descriptive questions and have been answered to a significant extent by many years of diverse study by amateur and professional alike. The next question has, again to some extent, been answered in the same way: where do they live? To be more explicit, over many years the accumulated, shared knowledge of many people has given us a very good idea of what species are found in Britain, how rare each one is, where it can be found in the country, and in what habitats. In a similar way something has been discovered about their ecology and behaviour: what they eat, where they roost in summer and winter. In addition to being fascinating, this information has guided conservation work. That said, the information available for most species is far from complete.

Quantitative, reproducible and tailor-made approaches are needed to answer many other questions. What habitats and roosting sites are most important to a species? How big or small is the total population? Are their numbers increasing or decreasing? It is surprising how often in science our hunches prove to be wrong. The only way to be sure about many things is to measure them, and this is what survey work does. To find out which habitats are most important to bats, we go out and systematically count bats (or some measure of their activity), using methods that allow us to compare results from different habitats, in different parts of the country, and obtained by different people. Through the early 1990s, hundreds of volunteers around the country took part in the National Bat Habitat Survey. On several nights each summer we all went for a walk in the dark. We set off at the same time, walked the same distance and counted bats in the same way, with bat detectors, marking each bat heard on maps that were distributed to us from a central co-ordinator at the Bat Conservation Trust, to whom we returned our completed maps. At the end of the study hundreds of 1-kilometre squares were analysed. Using nationally accepted classifications for different habitats, the results were analysed. The study was able to show the relative importance of each habitat to bats and the distribution of bats around the country. Bats showed a clear preference for woodland and water, and the survey also demonstrated just how important linear habitat features such as hedgerows were in maintaining habitat continuity across a wide range of habitat types.

In 1995 the National Bat Monitoring Programme was set in motion. This was a five-year project funded by DETR, run by the Bat Conservation Trust and supported by government conservation agencies. Its aims were to evaluate monitoring methods and provide baseline data for improved population estimates and distribution maps. It collected data on eight of the most easily identifiable species, using at least two of the following monitoring methods for each: bat detector surveys of foraging bats, counts at summer nursery roosts and counts at winter hibernation sites. Volunteers were again used, but with varying degrees of skill and experience, depending upon the method being used. The results of the project have been published in very comprehensive report available from the BCT. The BCT has also recently published a distribution atlas, based on records from the last 20 years.

Smaller-scale surveys are increasingly common. Bat detector surveys have been used to establish how particular species use habitat on a smaller scale, identifying key features too subtle for the one-kilometre land class system used, for example, by the National Bat Habitat survey. They can also be used as preliminary surveys of well-defined sites, such as nature reserves, parks, valleys and woods, and as part of environmental impact assessments. Bat detector surveys do have some serious limitations, discussed in the next chapter, so other methods must often be used alongside them. In many instances it is necessary to catch bats to identify them. Some species, such as long-eared bats, have almost inaudible echolocation calls, and others, including most *Myotis* species, are difficult or nearly impossible to identify from their echolocation calls. Roosts can be surveyed by repeated counts of emerging bats. This method also has its problems, but can provide information on population trends or on the effects of changes to the roost, such as recent building work. Hibernating bats can also be counted, again to look at population trends or perhaps to assess the effects of disturbance to particular sites. Some surveys have targeted particular types of buildings, such as churches and barns, and may have an important public relations function as well as a practical conservation element. Several major surveys of bridges have also been carried out to assess their importance to bats in different areas.

Practical conservation

Protection of key sites

Under UK legislation, 43 SSSIs in England, 26 in Wales and one ASSI in Northern Ireland have been designated specifically to protect key bat sites. All are roost sites. A further 77 sites have a listed bat element and others have varying relevance to bats. Proposed SACs include 12 bat sites. These are also based primarily around roost sites. Almost without exception both the SACs and SSSIs are in the south of England and Wales, reflecting the distribution of the countries' rarer bats, in particular horseshoes, barbastelles and Bechstein's.

The Vincent Wildlife Trust has about 40 reserves, primarily to protect horseshoe bats, but also Bechstein's, Natterer's and Leisler's bats. As the biggest

Fig. 6.2 Horner Wood, Somerset. Part of the National Trust's Holnicote Estate, and home to at least 14 species of bat, including a well-studied population of barbastelle bats.

owner of prime real estate in the country, the National Trust is responsible for many sites important to bats: they are an integral element of the management plans of many properties. Horner Wood in Somerset (Fig. 6.2) is one example. Fourteen bat species have been recorded in the wood and surrounding Holnicote Estate. One attraction is the wealth of roosts provided by the many old trees. There are some very unlikely reserves too. A dead tree in Cambridgeshire, home to a colony of noctule bats, was threatened with removal. The tree and a very small plot of land around it were bought and declared a local nature reserve.

The work at many of these sites involves not only protection and monitoring, but also management and restoration of both roost sites and feeding habitat. It may even include the creation of new roost sites and habitat. Several horse-shoe bat roosts have undergone extensive restoration, to preserve the build-ings themselves and to enhance their value to the bats. This may even include 'central heating' in nursery roosts. Habitat restoration is similar to that carried out for much of our wildlife: reduced grazing pressure and chemical use, restoration of hedgerows and woodlands and the creation of wetland habitats. Other measures are specifically bat-oriented, such as providing sheltered com-muting routes from the roost.

Creating roosts

From preserving, restoring and modifying roosts it is a small step to modifying structures in the hope of attracting bats, and then to building them from scratch. Pioneering steps in bat roost creation were made decades ago in the United States with the building of wooden towers to attract large colonies of Mexican free-tailed bats, a natural pest control agent and supplier of guano for fertiliser. The idea was a good one, but the design of the bat towers less so, since they attracted few bats. Recent attempts, led by Bat Conservation International (BCI), have been more successful.

Bat boxes

In European forests, a wide variety of small roosting boxes or 'bat boxes' have been developed over the last 40 years, to replace the natural roosts lost to forestry operations. These are generally small structures, made of wood, con-crete or woodcrete, attached to the main trunk of trees. They range from sim-ple boxes with slit entrances to elaborate structures with double entrances to limit air flow (Fig. 6.3). There are many thousands of bat boxes in place in Britain alone. Although some are single boxes put in place by interested indi-viduals, most are part of large schemes in woods or forestry plantations, initi-ated by the Forestry Commission, bat groups or other conservation organisa-tions. Occupancy rates vary from 10 to almost 100 per cent, although even occupied boxes may only be used for a few weeks each year. It is hard to be sure which designs are best, since surveys are often not systematic and few trials have been specifically designed to test one design against another. However, there is a little evidence to suggest that bats prefer woodcrete boxes to stan-dard wooden designs and they are also the most durable. Some species, such as brown long-eared, Daubenton's and Natterer's bats, and even the rare Bechstein's bat, will use them as nursery roosts. Pipistrelles more frequently use them as mating roosts in the autumn. Other *Myotis* species, noctules and Leisler's also make use of them.

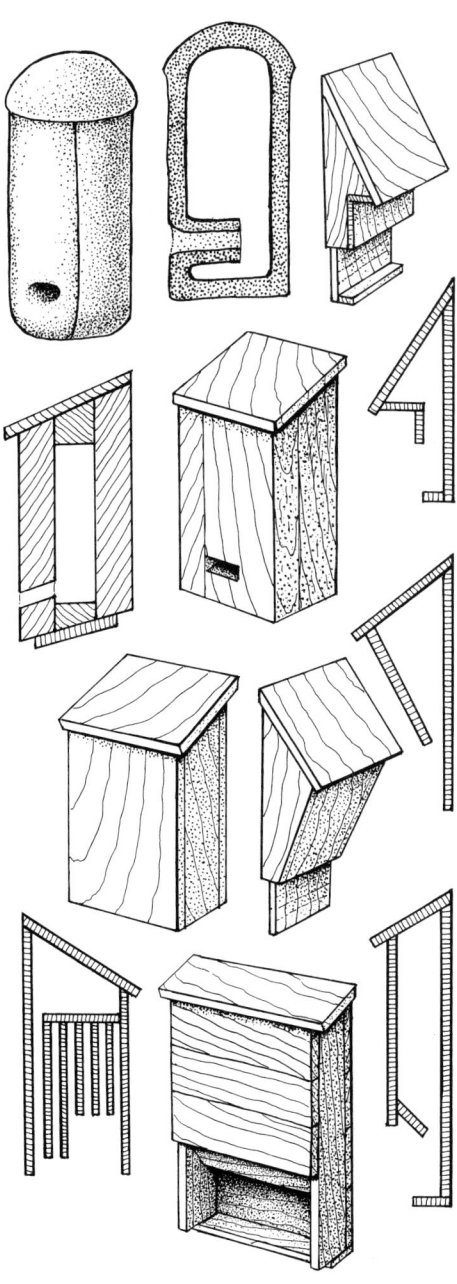

Fig. 6.3 Small bat box designs.

Bat boxes are often put forward as a partial remedy for habitat loss due to development. This is a very poor remedy: bat boxes are not a long-term alternative to natural tree roosts, and should be seen only as a stopgap in existing degraded woodland or in forestry plantations. As yet, there is no evidence to suggest they increase or even help maintain local bat populations. When more sympathetic woodland management practices have been operating for sufficient time, natural roosts in old and damaged trees will hopefully be abundant enough to make bat boxes redundant. They are labour intensive to build and maintain (few last as long as ten years) and some designs appear less than ideal. However, large schemes can be excellent public relations exercises; showing members of the public the bats as part of routine inspection causes little disturbance to the bats and can create considerable interest. They are also useful research tools: it is often easier to study bats using boxes than those in inaccessible tree holes.

Over the years a number of quite large boxes have been attached to trees, either as potential nursery roosts or as hibernacula. There is very little information on their occupancy rates, but some appear to have been better received by bats than small boxes. Boxes built to designs published by BCI appear to have been very successful in the United States, and are being built in Britain. Some multi-chambered BCI-designed boxes built by Forest

Fig. 6.4 A multi-chambered bat box built by Forest Enterprise in the North York Moors.

Enterprise (Forestry Commission) in the North York Moors Forest Park (Fig. 6.4) have a very much higher occupancy rate than adjacent 'conventional' designs. They are basically a variable number of narrow chambers stacked together, each just wide enough for a bat. The upper sections of the outer chambers can be filled with insulating material (as can the top), to retain warm air in the central chambers.

Hibernacula

The most ambitious artificial bat roosts constructed have been hibernacula (Fig. 6.5). We are moderately well informed about what makes a good hibernaculum and the science of building hibernacula with a wide range of microclimates is improving. Some of the more than 20 purpose-built hibernacula in Britain should therefore be very attractive to bats. Unfortunately, with few exceptions, the bats appear to think otherwise: most have been used by few bats and some have yet to record a single bat. It may be that enough is not yet known about the conditions required, but it is also possible that it takes bats many years to find and make use of them. Many of Britain's known hibernacula are man-made, but those attracting the most bats have generally been in existence for several decades at least. Building hibernacula, like planting trees, requires patience. One way to (perhaps) get quicker results is to

Fig. 6.5 An artificial hibernaculum built by Forest Enterprise in the North York Moors. Photograph: Charles Critchley.

improve structures already being used by bats, but it is essential to be certain that any changes really will be improvements. The first step is often to restrict access. Disused railway and canal tunnels, ice houses, lime kilns and cellars are all used by hibernating bats, but can all be subject to considerable disturbance and sometimes deliberate vandalism that can quickly drive bats away. Restricting access by gating the entrance can lead to dramatic increases in the number of bats using such sites. However, even this must be done with great care. The best hibernacula are frequently those with an air flow pattern which gives a broad range of microclimates. As the complexity of a hibernaculum increases, for example, by having passages and entrances at different elevations, the dynamics of the air flowing through them also becomes more complex. The addition of a gate changes the structure of the entrance, and can change these dynamics to the detriment of the bats. The best gates are frequently open grilles that do little to alter air flow and allow easy access to flying bats (Fig. 6.5). However, in some cases, restricting the air flow can improve a hibernaculum: if flow between the inside and outside is restricted, sites that are too cold can be made a little warmer. For example, after a railway tunnel in Wiltshire was partially sealed and grilles added, the number of bats using the tunnel increased significantly. Every site needs to be examined carefully and a unique solution devised.

Some novel structures have also been converted into hibernacula. Concrete military bunkers are still common in many parts of the country, and some are in suitably undisturbed locations. Temperature, humidity and light can be stabilised at the right level by partially blocking and adding grilles to the doors and by blocking gun ports and windows. This can all be done quickly and inexpensively.

The value of these artificial hibernacula can be increased by providing suitable internal roosting sites. Most British bats prefer to hide in crevices rather than roost in the open. The list of ways to provide these is endless, and many solutions are inexpensive and simple, such as the attachment of battens and boards to walls, behind which bats can crawl. Rubble can be piled against walls and old ceramic pipes stacked in corners. Purpose-built bat bricks are available, with numerous bat-sized cavities in them.

Summer roosts

Summer roosts can also be enhanced by the provision of roosting cavities within larger structures. Several bridges now have purpose-built roosts within their structure. Some very important bat roosts have been lost in the past due to demolition or major refurbishment. In such cases the only way to save such a roost may be to provide an alternative site. The Countryside Council for Wales commissioned a design for a lesser horseshoe bat roost to guard against the potential loss of an important roost. In the event that the original roost had to be destroyed, the new roost was to be built nearby in the hope that the bats would adopt it. This species has complex requirements, such as flyways into and through the structure direct to a large roosting chamber, a range of microclimates and a preference for locations where summer and winter roosting sites are all close to each other. Whether such an artificial roost will be adopted by bats cannot be said for certain: no project this ambitious has ever been attempted. To maximise the chance of success, an ambitious building was designed, based on all the available information on the requirements of lesser horseshoe bats (Fig. 6.6). The roof has different microclimatic zones and several alternative entrances. Below it are several cooler rooms and lower still, in the sloping ground on which the roost is built, is a cellar suitable for hibernation. As yet, there has been no need to build it. However, there is much to be said for building such a structure now, within the home range of a substantial colony, before it is needed in earnest. It would then be possible to experiment with and perfect the design for future use at other sites. It would also be an invaluable research tool, since it is designed so that the roost may be monitored in every conceivable way. Perhaps an enterprising company will step forward and fund this flagship project.

The variety of artificial roosts is limited only by imagination and the resources available. The various organisations listed at the end of the book provide detailed practical guidance.

Habitat conservation

I know of no cases of habitat restoration or conservation specifically for bats, unless related to key roost sites. However, large- and small-scale studies aimed at determining the habitat characteristics important to bats have been used in the preparation of various species action plans and management guidelines. These are often incorporated into local action plans that provide a blueprint

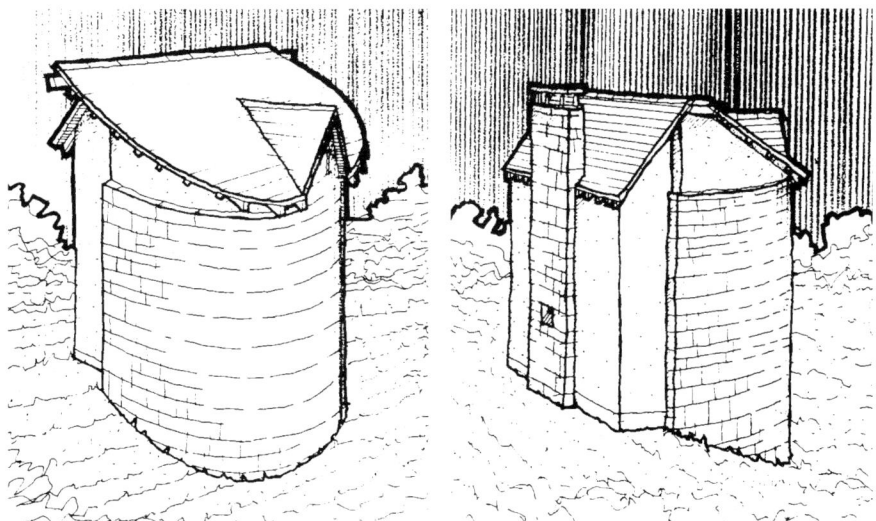

Fig. 6.6 The artificial lesser horseshoe bat roost, commissioned by the Countryside Council for Wales and designed with the help of Peter Dickeson Associates, Leeds.

for the conservation of a wide range of species, often in a variety of habitats. To provide an idea of how these work the following extract is a distillation of some of the important elements from management guidelines prepared for Daubenton's bats.

Management for Daubenton's Bat

Because there is evidence for sexual segregation of roosting and foraging bats, management should be at the catchment level to ensure protection of a viable, reproducing population. Habitat management at this scale would also benefit other long-lived and mobile species, such as otters.

Woodland and vegetation

Riparian woodland corridors should be created or maintained to encourage a diversity of insect species. Trees should be maintained or planted on both banks of rivers and static water bodies. To avoid excessive shading, planting density should be low and variable, with frequent small gaps. A mixture of species of different heights and structure should be used. Native species and/or those historically appropriate to the locality should be planted whenever possible. Linear landscape elements along which the bats can commute are also important. Hedgerows should therefore be encouraged as field boundaries, and trees maintained and/or planted on small feeder streams to provide continuity of riparian woodland between roosting and prime feeding sites. This woodland should contain trees that can be used as day/night roosts. Veteran trees should be retained, and possibly managed. Designation of tranquil areas of river banks, canals and lakes could provide havens for otters, roost sites for bats and feeding and resting areas for other wildlife. These areas should not be grazed by livestock, and there should be a general policy of non-intervention.

Maintaining, and where possible enhancing, riparian woodland would serve to buffer water bodies from the effects of agrochemical spray and other pollutants such as manure, slurry and effluent.

Foraging sites

The density and distribution of Daubenton's bat, particularly of nursery roosts, may be limited by the area of suitable water over which bats can forage. Smooth pools should be encouraged on stretches of river with surrounding riparian vegetation, particularly where trees are present on one or both sides of the river, since these are primary foraging sites. In lakes (and large ponds) sheltered bays in shallow areas, in which bats can feed, can be created by tree planting. South-facing locations are warmer and more productive. The creation or re-creation of water bodies (for example, restoration of large ponds and canal systems) with appropriate riparian vegetation, should be encouraged.

Areas of riparian marginal (woody) vegetation should be retained during river engineering works in order to maintain continuity of foraging habitat, particularly if the engineering work covers long stretches of river channel.

Roost sites

Woodland areas close to rivers and other water bodies should be maintained or created, and old trees preserved for potential roost sites. Renovation of bridges and buildings close to rivers should be undertaken carefully to prevent the exclusion or trapping of bats, through blocking of the roost site entrances and destruction of roosts. Potential hibernacula such as caves, mines and tunnels should be left undisturbed where possible. Entrances to such sites may be restricted through the use of grilles through which bats can pass.

Roost site creation in bridges and trees may be useful in situations where they may be a limiting factor. Specially designed roosts may be attached to bridges: the Bat Conservation Trust may be able to offer advice. The use of cordite and other explosives by the National Trust has shown potential in reducing top heaviness of trees and encouraging fungal entry and the formation of holes that may serve as roosts.

Most of these recommendations would benefit most of our native species, and there is some redundancy of effort and paperwork in the preparation of such guidelines for each separate species. However, this is unavoidable and often beneficial. Assumptions about what is good for different species must never be made: proper scientific survey is essential and often uncovers unexpected and critical requirements. Given the highly complex and often fragmented nature of our conservation network, with the expertise of many groups being limited to certain areas, it is also important to have clear and specific guidelines for different habitats and species.

Advice and consultation

Bat conservation frequently involves many individuals and organisations in the work itself. Much can also be achieved through the conservation equivalent of preventative medicine: it is better to anticipate and circumvent a problem than to have to cure it later. With this in mind, considerable effort goes into raising awareness about bats in government and commercial sectors. It is an unfortunate fact that even government departments can be unaware of their obligations or take them too lightly, and some avoidable incidents have resulted from poor communication. Having raised awareness, clear, practical advice can then given to ensure good practice in those activities that affect bats. The decisions made by central and local government, planners and architects have an enormous impact on both our built and rural environment. They must be aware not only of their legal obligations and the potential for damage, but of the opportunities to do good and to get support for good work. Local authorities and the Highways Agency are responsible for repairing bridges, work that can enhance as well as damage and destroy roosts. Throughout much of the country there exist channels of communications between these bodies, statutory conservation agencies and bat conservationists. The building industry can save as well as destroy roosts in buildings, as can firms treating timber for woodworm. Booklets are available and courses run to encourage good practice. There is an extensive network of organisations and incentive schemes to encourage farmers to manage their farms for the benefit of wildlife. There are even links with church authorities, custodians to many bat roosts. The Bat Conservation Trust can advise on all these aspects of conservation.

People who use the countryside for recreation also need to be aware of the potentially destructive nature of their activities. The days when cave life and even the caves themselves were damaged by visitors and entrepreneurs are largely over, but there is still a need for guidelines. The use of acetylene lamps by cavers should be discouraged in many caves, and winter visits to others kept to a minimum or discouraged altogether. Cavers should know not to disturb the bats they see, but to report their sightings. In recent years relations between cavers and biologists have been very good and there are many people who have an interest in both activities and can see both sides of the story. Anglers have long been asked not to discard hooks and line since they can ensnare wild animals. It can be a particular problem to bats, since a trout fly hanging over a river looks like food. I know of four recent fatalities on a single stretch of river. A ringed Daubenton's bat, followed for four years at one of our study sites, died this way. It is unlikely that more than a tiny fraction of such casualties are recovered, so the scale of the problem is unknown and the need to bring the matter to anglers' attention will never go away as new generations of anglers are born.

Finally, other conservation professionals need to be part of this network, so that the needs of bats are considered alongside those of better known or more 'charismatic' animals. As in many other professions, individual biologists have to become more and more specialised if they are to keep pace with an ever-growing knowledge base. Good conservation practice therefore relies on good teamwork, since no individual can be an expert in all the essential fields. In the past, despite their prominent place in Britain's vertebrate fauna, bats have often been forgotten or considered peripheral, but they now have a growing presence in the minds of the public and professionals.

Scientific research

The section on survey and monitoring considered relatively easy questions (at least conceptually, if not in practice) such as how many bats there are in Britain and where they are. This section addresses the frequently more difficult *why* questions. Why do we have as many or as few bats of a particular species? Why does a particular species prefer a particular habitat or roost? Why do different species have different social structures? Many of these questions are of real importance to conservationists. If we know why a particular habitat is important to a rare bat then it can be protected or even enhanced through active management. If we do not know, management is just guesswork. If we know what features of a roost site are most important, we are better able to provide artificial alternatives when natural sites are scarce, or a particular colony is under threat. Sound ecological knowledge leads to more informed, and hence more effective, conservation management decisions. Equally important is that these questions and answers are simply interesting. Many biologists are driven first and foremost by a desire to know how the natural world functions. Increasingly, as the threats to biodiversity become greater, this is accompanied by an equally strong desire to conserve. By passing on acquired knowledge to people from other walks of life biologists can, in turn, promote interest, fascination and the need for conservation.

Many questions come under the banner of studies that define a species' interactions with its environment. These have an obvious and direct relevance to habitat management and conservation. Why, for example, do Daubenton's bats prefer to feed over still water? Is it because that is where insects are most common, or because that is where the bats can detect their prey more easily? If insects are more common over still water, why and how does this influence the way our rivers are managed? Trees are important to foraging Daubenton's bats. Is this because they provide shelter for insects, because they reduce wind-induced surface ripples, or because the trees provide food and homes to the insects? Is the species of tree important? Will other structures, such as tall river banks, do a similar job? Is surface vegetation a factor in foraging site selection? The list of questions related to feeding is almost endless, and a similar number can be posed about roosting preferences. Some will be suggested as possible projects in the next chapter.

Foraging and roosting behaviour also raise important issues. Why do many species change roost so frequently? Is it due to changes in microclimate, parasite explosions, changes in the local distribution of prey or a combination of several of these factors? If microclimate is the only factor, it must be ensured that a wide range of roost types is conserved. If food is the key, then the nature of the roosts may be less important. A single factor is rarely the key, but studies of this nature do provide crucial information. It is vital, for example, that horseshoe bats are able to fly unimpeded into the roost. Measures to protect or enhance roosts that do not take this into account are largely doomed to failure, even if all other requirements are catered for.

Population ecology and genetics are becoming important areas of study and are closely linked to the problem of habitat fragmentation. How big an area is needed to sustain a viable population of a species? Does this area depend upon habitat type and structure? If the key habitat type is fragmented, can individuals move between habitat 'islands'? Over what distances, and through what habitats, is transfer sufficient to sustain a healthy population? Can wildlife corridors be used to link fragments of habitat? The answers are related not only to what the habitat can provide in the way of food and shelter, but also to the social and genetic structure of the population. Populations remain healthy only if the gene pool is sufficiently large. Female bats are often faithful to the nursery roosts in which they were born, and genetic flux most commonly (but by no means always) comes from the dispersal of males between colonies. The health of the population will therefore depend not only upon the size and distribution of habitat patches, but also on social structure and the dispersal patterns of offspring. Is a given habitat patch big enough to hold viable populations of its own or must the population recruit individuals from other patches? As patch size decreases, and suitable patches become more remote from each other, local extinction becomes more likely, but different patterns in social segregation, mating and migration complicate the outcome.

Mating in bats can take place in summer roosts, autumn mating roosts, swarming sites or hibernacula. This involves varying degrees of dispersal from, in most species, largely single-sex summer roosts. Species that undertake significant autumn migrations to communal swarming and hibernation sites prior to mating may be less prone to genetic isolation than sedentary species.

Genetics aside, fragmentation presents other threats. Small, isolated populations, even if genetically healthy, are more prone to extinction by chance events such as freak weather conditions, disease, introduced predators or human pressure. Local extinction can nibble at the edges of a species' range. The more common British species are widespread and found in a range of habitats: habitat fragmentation is probably less important than the more straightforward decrease in total habitat area or widespread habitat degradation. However, fragmentation could be of real significance to some rarer British bats. Known colonies of Bechstein's and barbastelle bats are a long way from each other. Losses in one are unlikely to be made up by migration from another. Are some of these populations already so small or genetically isolated or both that they are already doomed to extinction, whatever the efforts to protect them? We simply do not yet know. Local extinction may be a factor in the contraction of the ranges of both species of horseshoe bat in the last 100 years.

Knowing how an animal functions on the inside is also interesting and important to conservation. Ecophysiology looks at the interrelations between the internal machinery of a bat and its environment. Perhaps the best example in bats is hibernation. Bats hibernate to save energy when food is scarce, but what environmental and physiological factors control the internal clock of a bat? What factors trigger the release of the hormones that send a bat into hibernation? Is it the relative length of day and night, which is highly unlikely to change, whatever we do to the planet, or is it temperature, which is probably changing in the face of atmospheric pollution? Will climate change alter not only when bats go into hibernation, but the day-to-day cycle of torpor and arousal? How much energy is used up every time a bat arouses due to a rise in temperature or the presence of a biologist or a caver? How often can it arouse before it no longer has the reserves to get it through the winter? How easily are bats aroused by the light, sound and touch of a biologist? Most of these questions have only been partially answered, but all have a bearing on bat conservation. Similarly important questions can be raised about reproduction, such as when are nursing mothers and offspring most vulnerable to food shortage and stress brought on by bad weather and disturbance?

Education

Education is of vital importance to conservation. Humanity's attitude to wildlife and wilderness is extremely varied. There will always be people who see no value at all in the environment. Until quite recently, bats had a particular problem, in that if they had an image at all in the minds of many people, it was a poor one based on ignorance and superstition. I have even encountered this attitude amongst naturalists and professional biologists. Bats were often perceived as creepy, dangerous vermin, or at best inconsequential and insignificant oddities. Wanting to save the whale, the tiger or the panda was obviously a good thing. Wanting to save bats was just a little odd.

Fortunately, attitudes have changed a good deal in the last 20 years. Most people are now at least aware that there are bats in Britain, that they are not dangerous and that they have some measure of legal protection because they are becoming scarce. Whilst they may not be to everyone's taste, bats are seen as less of a threat. More encouraging is their rapidly growing positive image. A source of great satisfaction to many bat conservationists is the ease with which opinions can be changed. The level of ignorance can be so great that to merely show someone a bat can completely reverse their opinion of them. Many people are unaware that they can look so obviously harmless and so appealing. Having overcome this major hurdle with relative ease, the real process of education can begin. This too can be a straightforward task since bats excite curiosity, even if it is initially tinged with fear or revulsion: attracting an audience to a bat talk is not difficult. Information on bats with which to fascinate, entertain and inform an audience is there in abundance. We may know only a little of what there is to know, but that is still quite a lot, and delivered with enthusiasm it rarely fails to enthral.

The more people there are who understand bats the better, since everyone can have their experience enriched by a greater knowledge of the natural world, and all are potential allies and messengers. A lively public lecture can bring in a surprising (and sometimes daunting) number of invitations for more. To my more obvious audiences from schools, naturalist organisations and retirement clubs, I can add the rather less obvious Institute of Physics, Royal Air Force and aeronautical engineers, to name just a few.

Children and teachers

There is no better place to start than with children, either through schools or youth organisations. Catch them when they are young and you influence them for life. Children often have a great influence over the attitudes of their parents. Talks, bat walks, roost emergence counts, quizzes and biology projects are the most obvious events, but all sorts of ideas have been tried.

Professional organisations

In addition to informing professional bodies of their legal obligations, they can be helped to fulfil these and to make a positive contribution to bat conservation. Even a general talk on bats to planners, architects, environmental health agencies, builders, timber treatment and pest control companies can help by raising awareness. Specific courses can take this a step further by showing how to recognise that bats are using a building and by giving specific guidelines on what action to take if such evidence is found. Architects and planners can make a positive contribution by accommodating existing colonies and even encouraging new colonies in new or renovated buildings. In a similar way, those organisations responsible for the management of waterways, woodland, farmland and other habitats can learn from specific courses.

The government

Local government also needs to be kept informed of its obligations. Although the situation has improved enormously in recent years, mistakes are still made when improper advice is given, communications break down, or individuals simply do not know what to do. Local government can influence policy, but national government determines it and provides the framework and resources for nature conservation. Without constant and determined pressure, governments will often give nature conservation, particularly that of specific groups of animals and plants, a low priority. Taking the cynical view, many politicians need to be persuaded that a significant number of voters really care about a conservation issue. An important task of bat conservationists is to keep bats on the agenda with the rest of our biodiversity. Biologists may stress the need to see the big picture and view the natural world as a collection of interdependent organisms and ecosystems. However, legislation and funding are frequently targeted at specific species, habitats and locations: bats, like all other groups, need their champions.

The future

Although most of the individual actions that now change our environment and affect our bats are minor, it is this slow but steady loss of habitat that poses the greatest threat to bats and other wildlife. Education needs to be a widespread and continuing process, targeted at all organisations and individuals that can make a difference. Existing knowledge needs constant reinforcement and new knowledge needs to be disseminated. Children grow up, builders retire, governments change and new audiences arise all the time. Resources will always be limited and priorities therefore need to be established, but opportunities should not be passed over.

There is now strong legislation to protect bats and their roosts, but its implementation remains flawed. Prosecution can sometimes be counterproductive, in that it can turn opinion against bats. However, even with this provision, it has proven too easy in some cases to avoid prosecution for the destruction of bats and roosts, or for trivial penalties to be imposed. This can only undermine the effectiveness of the legislation. The habitat bats need is even less secure, because it is less easily defined and protected, and is subject to a greater variety of pressures. This is probably now the most important topic for future legislation, and will be tied to the conservation of habitats important to more than just bats.

Nature conservation may feature in speeches and policies, but the resources allocated to it by government have always been small: both practical conservation and conservation research are poorly funded. The CROW Act (p. 142) is a good step forward, but it will be no more than paper unless resources follow it.

To conserve our bats we need to know them – there is a pressing need for good survey and research, and there is much the amateur naturalist can contribute. Even the simplest research projects can give a focus and a meaning to natural history, in addition to assisting conservation efforts. The next chapter will look at just what the amateur naturalist can do.

7

Watching and Studying British Bats

For many bird watchers it is enough simply to identify birds or to watch them for pleasure, without any other purpose. This approach can satisfy many bat watchers too, but it is less likely to give lasting satisfaction: there are too few species and our views of them are all too fleeting. Identification is certainly a challenge, but alone offers limited opportunities. The enthusiastic bat watcher is usually attracted to more ambitious activities, as is the committed birder. This chapter covers the practicalities of bat watching: where to do it, how to do it, what equipment to use and how to give meaning to these activities for those who want to want to take it further.

There is much that the amateur naturalist can do to get more satisfaction from their activities and contribute to the scientific understanding of bats. There is no easily drawn boundary between natural history and scientific research, and it can be argued that in many instances each loses something of its interest and value without the other. Both require knowledge, training and the acquisition of certain skills, as well as enthusiasm and effort. There is no reason why amateurs cannot undertake research if they have the right training. So, even if you are not scientifically trained, or your scientific skills are rusty, do not let this put you off – get involved. The growth and organisation of the bat conservation movement, and the development of ever cheaper equipment to study bats, have made many things possible.

The practical discussion that follows will be confined largely to those activities that can be carried out without disturbing or catching bats and that therefore do not require a licence. More advanced or licensed activities will also be covered, but in less detail. If you do wish to try your hand at bat research, your local bat group will know where the bats are and may have some suggestions. The Joint Nature Conservation Council's *Bat Workers' Manual* is full of valuable information and will help you find a trainer when you get into more challenging studies. Any activity that involves disturbance or capture is illegal without a licence issued by the appropriate government conservation agency: English Nature, Scottish Natural Heritage or the Countryside Council for Wales. To get this licence you must train with a licensed trainer. You have to have a good understanding of bat biology and the laws protecting bats and be able to communicate this information effectively to the public. You must be competent in those practical skills you are licensed for and understand the implications of your work to the bats themselves. The licensing system is flexible, allowing you to be trained and licensed only for those activities you wish to carry out, so training can be relatively quick and efficient. The BCT, local bat groups and some other organisations run short training courses, to quickly bring larger groups up to speed on basic theory and practice. Courses are also run on specific aspects of bat work.

Equipment and techniques

Although natural history can be enjoyed with little in the way of equipment, being properly equipped can enhance your experience and make the impossible possible. Equipment is particularly important to those interested in bats, as it would be almost impossible to study many aspects of their life without it.

Bat detectors

Without any doubt, the most valuable piece of equipment is a bat detector. Small, mobile and nocturnal animals are not the easiest to study. Fortunately, bats echolocate, and being able to listen to, record and analyse their calls is the single, most effective way of finding out what species they are and what they are doing. It has the advantage that the bats can go about their business undisturbed, so you can study natural behaviour. Best of all, anyone can do it: you do not need a licence to study bats this way, although you will benefit greatly from reliable training.

Bat detectors work on several different principles, but they all have the basic function of making bats' ultrasonic echolocation calls audible to humans. With the more sophisticated (and regrettably more expensive) it is possible to record and then analyse the calls. Bat detectors are available in kit form for less than £30 if you have the basic skills needed to put them together: an easy day's work if you can handle a soldering iron and drill holes in a plastic case. Alternatively you can pay from £35 to £150 for a simple detector or a minimum of £250 for a more versatile unit. A top of the range research model can cost over £2,000.

Fig. 7.1 A selection of bat detectors.

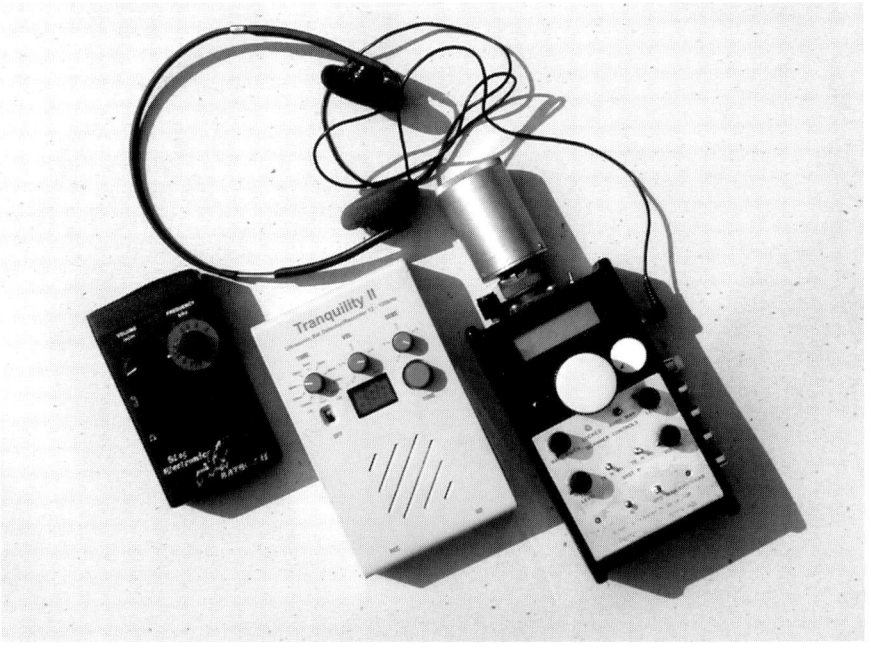

Heterodyne bat detectors

The most basic units, and the cheapest, are heterodyne detectors. In all types the echolocation calls are picked up by an ultrasonic microphone. In the detector's simplest form the signal from the microphone is mixed with that from a tuneable oscillator in the detector to bring the output down to a frequency humans can hear. Most detectors have a bandwidth of about 10 kHz (5 kHz either side of the tuned frequency) and will only produce an audible output when a bat's call is within this range. By rotating the tuning dial you can tune in to bats echolocating at different frequencies. Most modern detectors are in fact superheterodyne detectors, since they have two oscillator and amplifier stages. With two amplifiers boosting the signal, these detectors can have a high gain. The low bandwidth of the first amplifier lets little noise through, so excessive noise is not amplified along with the important signal, and the signal to noise ratio is high. The result is a sensitive detector with very little annoying hiss and crackle. Together with the low price, this makes heterodyne bat detectors an excellent general purpose tool, ideal for the beginner or those not too concerned with the finer points of identification and call analysis. They are also excellent to take on public bat walks, because they give clear, loud signals and a well-funded bat group can afford several to pass around.

Tuning a heterodyne detector until the most intense (loudest) signal is received tells you at approximately what frequency the echolocation call contains the most energy. This can, at the very least, enable bats to be separated into major groups, as shown in Figure 7.2. Bats that put out their loudest signals below 30 kHz are likely to be noctules, Leisler's or serotines. Pipistrelles, *Myotis* species, barbastelles and long-eared bats will be picked up best around 40–60 kHz, although long-eared bats will only be detected from very close range and are frequently overlooked. Greater horseshoe bats call at about 83 kHz, and lesser horseshoe bats at 105–110 kHz. However, there are many pitfalls and it is important to be guided by someone with real experience, and an honest appraisal of their own abilities, when learning how to distinguish the calls. Unreliable identification is of no value to anyone: better to admit to uncertainty than to guess.

So what are these pitfalls? Many bats pass so quickly that there is little time to tune in accurately. Variability in effective bandwidth of the detector and calibration errors add to the uncertainty. Most detectors tuned to 45 kHz will pick up strong signals between 35 and 55 kHz. If the detector is not properly calibrated a call at 35 kHz may really be at 25 kHz. Some manufacturers explain how to check the calibration and will tell you the effective

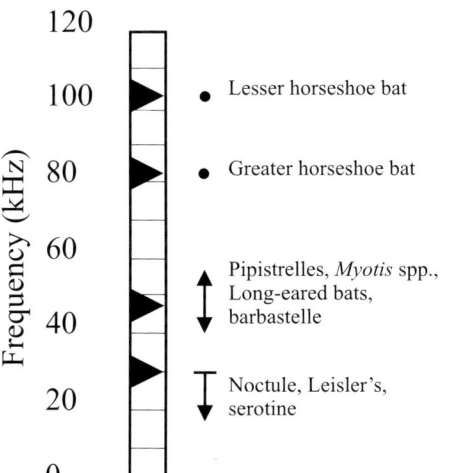

Fig. 7.2 The major groupings of bat species based on call frequency alone.

bandwidth. Even tuned to 45 kHz, a heterodyne detector can pick up the powerful noctule. The call frequencies of many species overlap and different individuals of the same species may call at slightly different frequencies. Finally, individual bats of some species can use different call structures at different times. Bearing these complications in mind, heterodyne detectors can still reveal some more subtle features. First of all, if a bat stays around long enough, you can find the highest and lowest detectable frequencies. This can help a little, but is subject to similar problems. Remember also that the higher frequency sounds suffer greater attenuation, so that the highest detectable frequency can vary enormously. Some guidelines are given in Table 7.1.

Table 7.1 Summary of high, low and most intense frequencies of bat calls.

	Call type	Average pulse duration (ms)	Start F (kHz)	End F (kHz)	F_{maxE} (kHz)
Greater horseshoe bat	CF-FM-CF	49(30-70)	69(64-78)	70(63-81)	82(81-84)
Lesser horseshoe bat	CF-FM-CF	46(20-60)	98(86-108)	96(84-110)	109(106-113)
Bechstein's bat	Broadband FM	2.5(1.6-3.0)	111(65-131)	34(28-39)	51(46-55)
				42(34-50)*	73(55-91)*
Natterer's bat	Broadband FM	2.3(0.5-5.3)	99(57-146)	23(15-47)	51(27-81)
			121(96-146)*	31(20-42)*	65(30-95)*
Daubenton's bat	Broadband FM	2.9(1.3-5.8)	81(49-110)	29(22-40)	46(30-55)
			88(78-98)*	36(25-47)*	55(47-63)*
Whiskered bat	Broadband FM	2.2(0.3-4.0)	80(56-102)	32(27-39)	48(39-65)
			103(79-127)*	40(30-50)*	58(43-73)
Brandt's bat	Broadband FM	3.1(1.5-5.0)	86(59-123)	34(27-42)	48(38-78)
			101(84-118)*	55(43-67)*	
Serotine	Narrowband FM-CF	5.2(1.6-11.7)	57(39-78)	28(22-32)	32(26-42)
Noctule	Narrowband FM-CF	19.6(11-34)	26(19-53)	18(15-24)	20(17-26)
Type 1 call*		11.5(6-17)*	47(30-64)*	26(22-30)*	27(22-33)*
Type 2 call*		13.1(8-18)*	31(20-42)*	22(18-26)*	22(15-29)*
Leisler's bat	Narrowband FM-CF	8.0(4-15)	44(25-80)	25(19-29)	28(22-35)
			59(32-86)*		
45 kHz pipistrelle	FM-CF	4.8(2.6-7.3)	71(54-119)	43(39-48)	46(42-51)
55 kHz pipistrelle	FM-CF	4.1(2.2-6.6)	82(65-114)	51(45-54)	54(49-58)
Nathusius' pipistrelle	FM-CF	6.12(3.0-7.7)	50(40-67)	37(35-38)	39(37-41)
Barbastelle(1 bat)	FM	4.4	46	28	37
Type 1 call*	FM	2.5(1.3-3.7)*	38(31-47)*	30(25-35)*	34(27-41)*
Type 2 call*	FM	3.8(2.3-5.3)*	46(41-51)*	31(23-39)*	39(32-46)*
Brown long-eared bat	Broadband FM	1.5(0.5-2.5)	61(43-100)	29(20-43)	43(30-75)
					52(26-78)*

Data from Vaughan et al. (1997). Figures in brackets are approximate upper and lower limits. Additional data * from Parsons & Jones (2000) are given when the two sources vary significantly. The ranges for these data are rough estimates (based on the standard errors and number of calls analysed quoted in the paper, assuming normal distribution about the mean).(NB. F_{maxE} can only be determined with a time-expansion detector described later.)

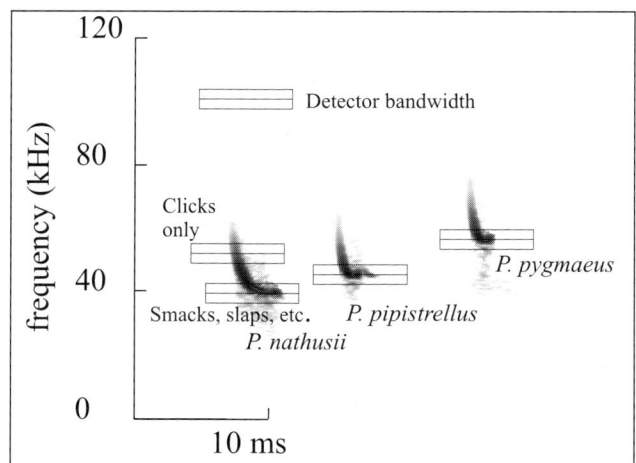

Fig. 7.3 The calls of the three pipistrelle species and how to tune in to them with a heterodyne detector.

Some species have calls with distinctive tonal qualities on a heterodyne detector. The long constant frequency components of horseshoe bat calls are heard as long, melodic warblings and cannot be mistaken for any other species. Pipistrelle calls, if you tune in to the quasi-CF or narrowband tails, are short melodic slaps, smacks, chips, chops or plops: it is not easy to describe them, but it does not take long to learn to recognise them. By tuning into the plops of the three pipistrelle species it is possible to separate these with a high degree of confidence. This is possible because there is minimal overlap between the frequency ranges of the narrowband components of their calls (Fig. 7.3). Accurate determination of the quasi-CF tail to the call requires that you tune the detector until the plops have their lowest pitch. The bars in the figure represent the bandwidth of typical detectors: tuned in to 45 kHz, most detectors will pick up strong signals between 42 and 48 kHz, so careful tuning is needed.

Similar chips, chops, plops and tocks, but detected at a lower frequency, are heard from the narrowband calls of noctules, Leisler's and serotines, but there is overlap in the frequencies of their calls and separating these species on this basis can be risky (Table 7.1). Noctules have the lowest frequency calls, with the tail of the call centred about 22–25 kHz. The tails of Leisler's and serotines are both centred at about 28 kHz: too close to allow identification on this basis alone. There are other clues to the identity of these three species. In the open, noctules often alternate narrowband and broadband calls, for reasons discussed in Chapter 4. With a little tuning they are heard as a distinctive chip-chop, chip-chop sound, at a low repetition rate. Repetition rate increases when the bats fly close to trees or the ground. Leisler's bats also use this alternating call, but less frequently: serotines do not.

Tuning into the steeper frequency modulated components of bat calls produces short, sharp clicks. Since the calls of *Myotis* and *Plecotus* species have no other components, this is all you will hear however much you tune your detector. Heterodyne detectors cannot be used to separate these species with an acceptable degree of confidence. However, if you can see that a bat is flying within a metre or two of you, but its call is barely audible, it is probably a long-eared bat. If it is feeding by flying very low over water, it is probably a Daubenton's bat.

Many species produce a variety of what are loosely called 'social' calls (p. 168). These are typically lower in frequency than echolocation calls and multisyllabic. Although distinctive it is rarely possible to pin them down to a species using a heterodyne bat detector.

Besides those already discussed, there are other disadvantages to heterodyne detectors. First of all, you can only listen in on one frequency at a time, so unless you constantly scan the frequencies, bats may pass undetected. The major disadvantage is that the very process that makes heterodyne detectors so sensitive and noise-free also throws away most of the detailed structure of the call. Spectrographic analysis, as we will soon see, can be very revealing and very interesting, but it is simply not possible using heterodyne detectors.

Finally, a word of caution. Many people claim to be able to identify species, using heterodyne detectors, on the basis of very subtle differences in the tonal quality and rhythm of their echolocation calls, including such similar sounding species as Natterer's, Daubenton's, Brandt's and whiskered bats. On purely scientific grounds this has to be viewed with considerable scepticism. A method can only be valid if it can be verified. The claimed differences are so subtle, in the face of known variation within species, that they are often unverifiable. Whilst the technique may be valid in skilled hands, I have seen confidently announced identifications proven wrong when the bat in question has conveniently flown into a trap for inspection in the hand. A basic rule of good science is that you have to be able to give objective evidence in support of your conclusions.

Frequency division bat detectors

This method uses a device called a zero-crossing circuit to identify the fundamental frequency component of the echolocation call, which is then divided, typically ten-fold, to lower the frequency into the audible range. The major advantage is that these are broadband detectors, and are able to pick up species echolocating between 10 and 150 kHz or beyond, without constant tuning. Although all harmonic and intensity information is lost, frequency information is preserved, so basic spectrographic analysis is possible. A major disadvantage is that all sounds, including background noise, are amplified. Signal to noise ratios can be low and these detectors can be noisy and irritating, although this can be overcome with additional processing. There are systems available that enable you to generate spectrograms in the field based on frequency division techniques, and software is being developed that can identify bats based on this information. Whilst this works for some species, the approach has many critics (myself included), because so much of the information in the call is lost, and there is little need to persist in using the technique when a better one is available. Their one big advantage is that they do give 'real time' spectrograms.

Time-expansion bat detectors

Until recently this type of detector was so expensive that only professionals and wealthy or determined amateurs could justify the cost. As digital technology has become cheaper, time-expansion detectors have become more affordable. They are still the most expensive, but offer significant advantages. The signal from the ultrasonic microphone is captured by a fast analogue to digital (A/D) converter at very high sampling rates: fast enough to preserve the detail of even the highest frequency echolocation calls. This information is temporarily stored in a memory buffer. It can then be played back through a digital to analogue (D/A) converter at a slower rate, and thus at a frequency range low enough to be recorded onto an ordinary audio tape recorder. Commercial detectors slow the playback (time-expand) 10-, 20- or 32-fold. Ten-fold time expansion is the most useful for recording and subsequent analysis, but 20- and 32-times expansion can help you to hear differences in call frequency in the field.

Having downloaded calls to audio tapes they can be analysed on the spot if you have a laptop computer, or taken home for later analysis. For most purposes, inexpensive, portable cassette recorders and inexpensive audio tapes are perfectly adequate. Minidisk recorders are now being used by some bat workers. They have a very good frequency response, but the very process that makes it possible to squeeze so much sound onto these tiny disks is a potential source of error. A data compression system shrinks the files by throwing away sound components the human ear will not detect if other, more intense frequencies, will mask them. This could be disposing of critical information. However, in a quite detailed comparison of the same calls recorded onto both a minidisk player and a professional quality cassette recorder I detected no differences in call structure. This suggests that the changes are likely to be of concern only to scientists doing detailed analysis. Minidisk recorders are well worth considering for many tasks, since you can store several hours of recordings on a single disk and access particular tracks rapidly.

Time-expansion bat detectors are without doubt the best option for serious bat watchers. By preserving virtually all the frequency and intensity information contained in the calls, including that of any harmonics, they make detailed spectrographic analysis possible. This is the only reliable and verifiable way to identify some bat species and it reveals a wealth of detail for further study.

Spectrographic sound analysis

As visual animals, the best way for humans to study bat calls is to be able to see them. Until a decade or so ago, this really was the territory of the professional biologist, since both expertise and expensive equipment were necessary. The ever-decreasing cost and increasing power of the personal computer has changed all that. If you have even the cheapest time-expansion bat detector, a small tape recorder and a home computer, then you have all you need to start. First look at your tape recorder and see if it has input and output sockets: they usually fit 3-millimetre 'jack' plugs of the type that connect headphones to personal hi-fi equipment. On more sophisticated machines they may be available as separate line-in and line-out sockets, but microphone and headphone sockets will do. You will need a cable to connect your detector output to the line-in/microphone socket of the tape recorder to record your time-expanded signals, and later to connect the line-out/headphones socket of your tape recorder to your computer sound card's line-in socket. Most home computers are now sold as 'multimedia' machines: they have a sound card and loudspeakers. If yours does not, then, unless it is very old, you can buy and fit these inexpensively and easily. Finally, you need software to perform the analysis. Software written specifically for echolocation call analysis is available from the maker of the more 'professional' bat detectors. It is expensive, but worthwhile

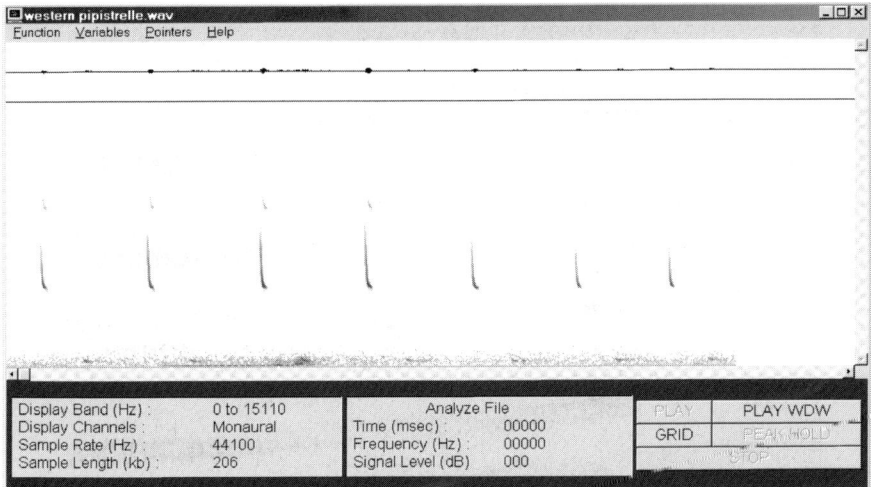

Fig. 7.4 Typical sonograms, as displayed by a software analysis package.

if you can afford it. Fortunately, excellent general purpose packages are available for little or no cost and many can be downloaded from the internet for trial periods. You may have to register as a user and pay a small fee to get a fully functional package, but this will usually entitle you to upgrades.

So what do these packages do? All of them allow you to input short segments of sound from your tape or minidisk recorder and store it in the computer, usually as a WAV or wave file. It is best to keep each file short (less than 30 seconds) unless you have a fast computer with plenty of memory, since these files can be large and unwieldy. However, bat calls are very short and even a few seconds of tape contains many calls. You can of course store many hundreds or even thousands of these calls. Some software allows you to download from your bat detector direct to a laptop computer. Having saved your file you can now view it in its most informative form, as a spectrogram or sonogram: a graph of sound frequency against time, as shown in Figure 7.4. Each echolocation call is revealed as short, discrete and with a characteristic form. The narrow window at the top is sound amplitude against time. You can see each call as a burst of energy.

You can position a cursor on different parts of the call to measure its bandwidth and start (F_{start}) and end (F_{end}) frequencies. You can describe its shape (for example, FM, FM/CF, FM/CF/FM), its duration and the time interval between calls. Remember that the signal is time-expanded: to get real durations and intervals you must divide by ten and to get real frequencies you must multiply by ten. Harmonic components may be present above the fundamental. With most software, different sound intensity levels are colour coded, so it is possible to determine, at least approximately, those frequencies containing the most energy (F_{maxE}). The latter can be determined more exactly with some of the software available by selecting a power spectrum or frequency analysis for a call, which reveal a graph of intensity against frequency. Figure 7.5 shows such a plot and the high energy peaks of the calls are clearly revealed. Again, since the calls were slowed down by a factor of ten, the frequency must be mul-

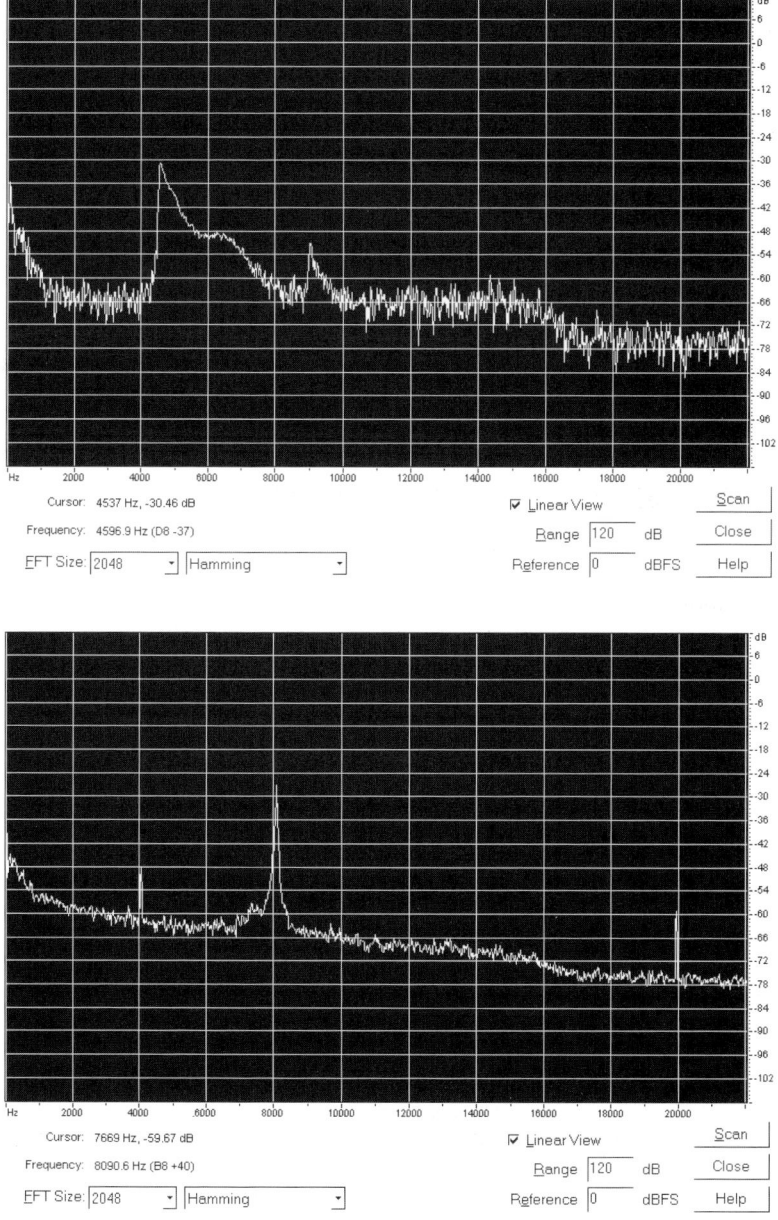

Fig. 7.5 Power spectra of FM/CF and FM/CF/FM bats.

tiplied by ten. Note the greater spread of energy across the frequency range in the pipistrelle call at the top, relative to the horseshoe bat call in the lower figure. Note also the peaks of the harmonics and the fact that most of the ener-

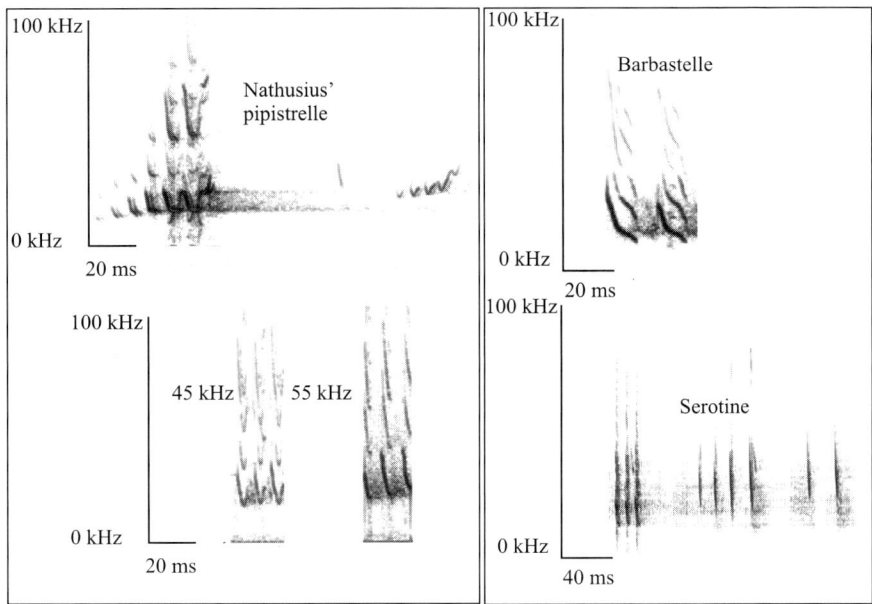

Fig. 7.6 Social calls of the three pipistrelle species.

Fig. 7.7 Social calls of barbastelle and serotine.

gy in the horseshoe bat's call is in the second harmonic, not the fundamental. This information is all you need to take identification almost as far as it is currently necessary to go: F_{start}, F_{end} and F_{maxE} are by far the most informative variables to measure and these are listed in Table 7.1 (p. 162). All the software is easy to use and the BCT runs bat detector workshops to teach the basics.

Other sounds that bats make

In addition to search phase echolocation calls, bats also produce feeding buzzes and a range of different social calls. Although these are often distinctive and informative when simply listened to in the field, they too can be subject to sound analysis to reveal new information. For example, it is possible to see not only the increase in call repetition rate during a feeding buzz, but also changes in bandwidth and harmonic composition and the terminal drop in call frequency characteristic of some species (Fig. 4.18). The complexity and variety of social calls can be seen. For example, the social call of Nathusius' pipistrelle has a component not seen in that of the other two species (Fig. 7.6), often making identification unambiguous. Analysis also gives important clues about what the bats are doing: social calls in late summer and autumn may be those of songflighting males. Social calls vary considerably, even within species, but most are broadband, terminate at low frequencies and have more than one syllable (Fig. 7.7).

Other field signs that aid identification

Although echolocation usually provides the most concrete information about a bat's identity, there are other clues. The most obvious are locality and rarity. A small bat emerging from a building or foraging in an urban environment is probably a 45 or 55 kHz pipistrelle, wherever in the country it is seen. A large bat in the north of England or Scotland is unlikely to be anything other than a noctule. I hardly need to labour this point, but its corollary is important. If you have a pipistrelle in the hand, do not assume it is a 45 or 55 kHz pipistrelle just because they are the most common: rarities are not found unless you look for them. Take the trouble to identify carefully any bat you are fortunate enough to get hold of, particularly if there is anything unusual about the circumstances.

Chapter 4 discussed why some bats emerge earlier than others (p. 102). Whatever the reasons, emergence time can be used to help reduce the options when trying to identify bats. A large bat flying in the open before dark is probably a noctule, but both Leisler's and serotines are also early risers. Small bats emerging from a roost before dark are most likely to be pipistrelles and least likely to be long-eared or Natterer's bats. Roost location is also a guide: large house-roosting bats may be serotines, since noctules use buildings infrequently. Small bats emerging from a tree hole are most probably a *Myotis* species: those emerging from a bridge Daubenton's bats. Bats flying in dense vegetation are most probably brown long-eared or Natterer's bats.

It is sometimes possible to get a glimpse of the shape of a bat in flight, particularly against a post-sunset glow. The ears of a long-eared bat, if you are lucky enough to get a glimpse, leave you in no doubt as to its identity. Serotines have shorter and broader wings than either noctules or Leisler's bats, a difference that is sometimes apparent if the bat flies close. Flight pattern can also help: Daubenton's is the only species that forages in very low, continuous sweeps over water. There are other features that help separate groups of bats, but beyond the obvious categories such as gleaners or open air foragers, most are unreliable.

Few of these clues are infallible and most, like many of the skills of natural historians, require practise and experience before they can be used with confidence. There are few short cuts to competence and to try to provide an exhausting list would be misleading. However, learning them is one of the pleasures of natural history. But remember, you cannot lay claim to a reliable record unless you have more tangible proof of identity.

Other useful equipment for studying active bats

Clearly, devices that help humans to see in the dark can be of real value in studying bats. As in other areas, technological improvements have brought a number of devices into the public domain. The most obvious are night scopes, once the preserve of the military, but now widely available (Fig. 7.8).

A night scope takes the very small amount of light available from the stars, moon or distant artificial source and amplifies it. Each photon of light that enters a photomultiplier tube generates a cascade of electrons that bounce off a screen coated with fluorescent material to create a bright image. If the ambient light is too low, invisible infrared light can be used to brighten the image. Night scopes are described as using first, second and third generation technology, with increasing, sensitivity, clarity and cost. An inexpensive first generation night scope, costing a little over £100, will give a weak image, best supplemented with an infrared light. The image will be sharp enough to be useful in the centre of the field of view, but will rapidly blur and distort towards the periphery. A second or third generation night scope will probably cost

Fig. 7.8 Night scope and 'nightshot' camcorder.

about £1,000 and frequently rather more, but could be worth the investment, giving a bright, sharp image to the edge of the field of view. Most night scopes magnify several fold and have a small field of view, making them very difficult to use. Flying bats are hard to follow and these models are useful primarily for watching bats during emergence. Heavier models will need a tripod. The most useful models, with little or no magnification and a wide field of view are expensive second and third generation night scopes. Many come with small infrared lights attached, but they are rarely powerful enough. The best way to get a broad, bright infrared light source is to fit a filter to an ordinary lamp. Manufactured infrared filters are not expensive, but you can also improvise with layers of red, transparent plastic film. Night scopes come as single tube telescopes and binoculars. Some binocular systems can be head or helmet mounted, leaving the arms free, but to wear the heavy, affordable models you need the neck muscles of an ox. After looking through a night scope for any length of time you may be as good as blind in that eye until it adapts to the dark again. This can be disturbing initially, but it is not hazardous.

In the last few years, domestic digital 8 camcorders have become a more affordable and lighter-weight alternative to night scopes, with the added advantages of remarkably good zoom lenses. The CCD (charge-coupled device) at the heart of a camcorder turns light falling onto a rectangular array of photodiodes into electrical signals that are then used to create the image. Most CCDs are very sensitive to infrared light. Because this increases the difficulty of creating a balanced colour image in daylight, an infrared filter is normally inserted in front of the CCD to eliminate the problem. Several manufacturers recently had the good sense to insert a switch so that the filter could be withdrawn, turning the camcorder into an effective night scope. The liquid crystal display (LCD) common on many models allows more than one person to view the image. Although the quality of the image on these displays is not so good, the image in the viewfinder and the recorded image are much better. The camcorders use an infrared light source in their auto-focusing system that can be surprisingly bright. Clip on lights are also available, but again nothing beats an improvised infrared lamp. Another benefit of a camcorder is that the sound system can be used to make observational notes or even to record echolocation calls through a time-expansion detector: both if the model is stereo. Camcorders are nowhere near as good as second generation night scopes, but many of the latter can be attached to the front of a camcorder.

Because roost temperature is so important to bats, temperature loggers often come in very useful, since they can be hidden and left unattended. A number of companies make durable, weatherproof loggers small enough to fit into the palm of your hand and capable of storing up to 16,000 data points over period as long as two or three years. They can then be downloaded to a personal computer. Both temperature and relative humidity can be stored by some models, although this increases the cost from £50–80 to up to £160. Similar loggers can be used as 'event loggers', recording the passing of an echolocating bat for example, with the addition of a simple heterodyne circuit and event switch. One of the bat detector manufacturers makes a 'bat logger' that can independently count bats when attached to a heterodyne detector. When used with a data logger it logs the time each bat passes, or counts the number of bats in a given time interval.

Restricted techniques that require a licence

Catching bats

Many important studies require capture and handling and these are potentially damaging to bats. With proper training and care they have a negligible effect on the bats. When using some techniques, constant re-appraisal of the value of the information in relation to the potential risks to the bats is essential. If there is any evidence that your activities could lead to increased mortality or decreased breeding success you must be confident that the scientific and conservation value of the activity merits this risk. If the risk is anything other than small it may be difficult to justify some activities at all, particularly in the case of rare species. A major problem is that the magnitude of the risk may be difficult or near impossible to determine once the bat has been released into the wild. Natural mortality rates are notoriously difficult to determine in the first place: assessing the contribution made by research doubly so. In the end we must rely on what little evidence we can collect together with common sense and a well-developed conscience. The best evidence comes from data on ringed and therefore identifiable bats recaptured over several or many years (p. 174).

Many techniques are used to catch bats, but here only the most common and widely used will be described. Bats emerging from the roost are best caught using static hand nets. The technique is safe and does not disturb those bats still in the roost. Foraging bats are usually caught in mist nets or harp traps. Mist nets are versatile, but they need to be put up correctly and considerable practise and skill are required to remove bats swiftly and safely. Supported at both ends by poles, mist nets are typically about three metres high and 6–30 metres long, although larger ones may be used in special circumstances. They are basically small-mesh nets made with fine nylon, hung in such a way that bats flying into them drop into pockets in the net, from which they are unable to escape. In competent hands a bat can be safely removed from the net in a minute or two at most, usually in just seconds. It is easier on both bat and

Fig. 7.9 Harp traps in
use outside a cave
entrance.

human if the nets are constantly manned and bats are removed as soon as they
enter the net. Mist nets should not be used when large numbers of bats are
expected, since you can be quickly overwhelmed. They can be set up close to
the ground or lifted high in the air on long poles, and several nets can be
arranged in patterns that maximise the chance of capture. Bats can and do
detect nets, often with frustrating ease, so they must be set up with some cun-
ning. Bats appear to fly much of the time on spatial memory, relying on a
detailed knowledge of familiar environments. They may concentrate more on
prey than on obstacles, so your aim is to catch bats by surprise. One net can
often be placed to catch bats that have turned to avoid other nets. Once bats
are aware of the presence of mist nets they avoid them with ease and capture
rate rapidly declines unless naïve bats arrive. For this reason it is usually best to
change the location of the nets each night.

Harp traps (Fig. 7.9) are less versatile, because they are smaller, rigid and
more cumbersome. However, they are easier to use and kinder to bats. In my
experience they are also more effective, typically catching more bats than an
adjacent mist net several times larger. In essence, a harp trap is two parallel
arrays of vertical monofilament nylon threads. The threads in each array are
just a few centimetres apart and the two arrays are a little further apart. The
threads of the second array lie in the gaps of the first. The frame holding these
arrays is light, but strong, since the threads must be held taut. The natural elas-
ticity of the nylon cushions the impact of bats, which drop into a catching bag
below the trap. This is designed to allow easy entry, but to prevent escape. The
catching area is typically about 3 metres high and 2.5 metres wide. Smaller and
larger traps are used for particular purposes. The trap stands on adjustable legs
or can be suspended. Harp traps are most effectively sited where bats are fun-
nelled into a narrow flyway: over a narrow, overhung stream, in a woodland
ride, or at a cave entrance. Bats can remain safely in the trap for some time,
but it is best to remove and release them as soon as possible. I usually stay with

a trap, rarely leave it for more than 30 minutes and never for more than one hour.

Handling bats

Although most species show little or no sign of stress (some struggle initially to escape) common sense suggests it is best to keep handling and containment to a minimum. We are generally interested in studying natural behaviour and must limit activities that might alter this in unknown ways.

There are many reasons why we need to catch bats if we are to learn more about them. Identification is impossible without close inspection. Even those bats that can be identified unambiguously by their echolocation calls must be captured initially if the call is to be matched with the bat. New research is constantly uncovering new species or altered distributions, that must be confirmed by capture. We need to be able to confirm sex, age or reproductive status to answers such questions as 'Is this a nursery roost?', and 'does this species breed here?' The collection of data on mortality and reproductive success requires bats to be captured at some stage in a study. The list is endless and more reasons will emerge throughout this chapter: just one more will be mentioned here. The best way of changing public perceptions about bats, dispelling some of the myths that have led to persecution and intolerance, is to show people real bats, close up: it rarely fails to impress.

Anyone handling bats should be protected against rabies. A rabies-like virus is present in bats in certain parts of Europe. A Daubenton's bat infected with the same virus was found on the south coast of England in 1996, and was probably a natural or human-assisted migrant. A second case was reported in 2002, again a Daubenton's bat. This individual was found in Lancashire and may have been a native bat. Many bats are tested and cleared every year and the most logical interpretation is that this virus may be present at very low levels in our native bats. Although the risk is probably extremely small, anyone finding a grounded or moribund bat should not touch it but contact English Nature or their equivalent or the local bat group. For people working on bats, the sensible precaution is to be vaccinated against rabies and take precautions to minimise the risk of being bitten. This is not the place to give detailed advice, which may change in the near future: consult your doctor and DEFRA for the latest procedures.

Marking and tracking bats

To answer many questions it is essential to be able to identify and follow individual bats and this means marking them in some way. Fur clipping can be used as a temporary mark: a neat trim leaves an obvious straight edge to the fur. By varying the number and location of the clips, a large number of bats can be given unique identifying marks. The scissors need to be very sharp, to avoid pulling the fur and to give a clean edge to the clip, but they should have blunt tips to avoid stabbing the bats. I have used fur clipping to make a preliminary estimate of the number of bats visiting a cave entrance, since I can take into account recaptures. One of the first things I learned was that very few bats are careless enough to get caught more than once in a night.

Many studies require that bats are identified with marks that persist for months or years, and clipped fur grows out quickly and is later moulted. Alloy rings are the most widely used long-term marking technique. There is evidence to show that early ringing studies led to significant increases in bat mortality. This was probably due to poor ring design and manufacture, poor fitting, and the fact that many early studies involved disturbing hibernating bats. Techniques have improved and ringing has also become much less common. By and large, bats are now ringed only as part of carefully planned studies with clearly defined objectives, where ringing can be shown to be essential. Ringing for any other reason should be discouraged. I am involved in several studies that involve ringing, some going back ten years or more: all have long-term goals that are constantly re-assessed. Once a bat is released we have no way of assessing the damage a ring might do unless the bat is recaptured. I have seen ringed bats that have been recaptured more than ten times, sometimes over as many years. Of the thousands seen, only a few have shown signs of injury. Although this is encouraging, I still do not ring bats without good reason.

The standard ring used is distributed by the Mammal Society. It is made of smooth aluminium and is engraved with a unique number and the words LOND. ZOO, to encourage the public to return rings found to London Zoo. The rings come in two sizes (Figs 7.10 and 7.11): a large ring for noctules, Leisler's, serotines and greater horseshoe bats and a smaller ring for all other species. Intermediate sizes were once used, but are now considered unnecessary. The ring fits over the forearm of the bat and is not quite fully closed, so that it can run freely up and down the arm, but without crossing the wrist and trapping fingers. These rings weigh only 50 and 100 milligrams. Coloured plastic rings are used in many countries, but rarely in Britain.

To determine where and how bats forage, a number of techniques are used. Small reflective disks can be temporarily glued to the wings of bats, or more permanently to rings. When illuminated by visible or infrared light, they can be seen over surprising distances. Reflective material comes in several colours, but they are not always easily distinguishable in practice. The best material is the Scotchlite used on reflective road signs. Better still are devices that generate their own light. Glass spheres that use harmless β-radiation to illuminate an internal phosphor coating have been used in the past, but I have not heard of them being used recently. Broken glass is a certainly a potential hazard with these. Cold chemical light sources such as Cyalume are a better, but short-

Fig. 7.10 Bat rings, radio transmitters and a Cyalume light tag.

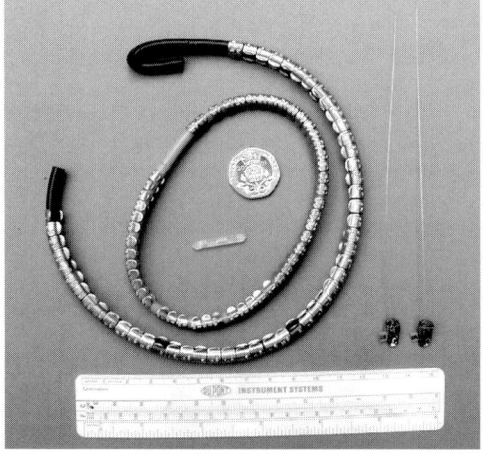

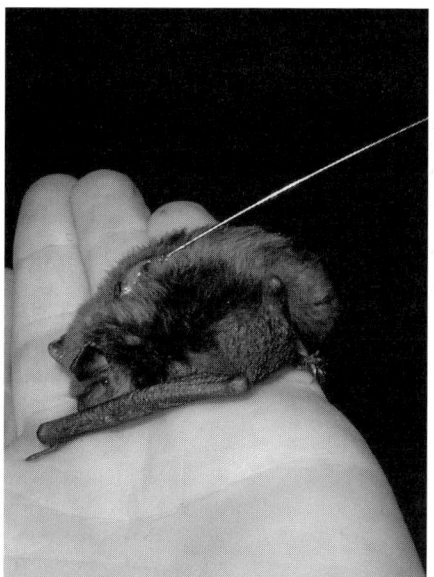 

Fig. 7.11 Daubenton's bat fitted with a radio transmitter and a ringed Natterer's bat.

lived, option. This is now conveniently available in small plastic cylinders (20 x 3 millimetres, 130 milligrams (Fig. 7.10)). Bending the cylinder cracks an internal glass tube, allowing two chemicals to mix and giving a strong green light for up to 12 hours. Glued to the back of a bat they can be seen 20 metres or more away in complete darkness and over much greater distances with night scopes. If glued lightly to the fur bats will easily groom them off on returning to the roost. They have the advantage of being inexpensive, but they last only one night. If following a bat that has a large foraging area or unpredictable foraging patterns, you may never find it again once lost.

If you are determined to track bats, then the only realistic option is radio-tracking (also known as radiotelemetry). This involves gluing a small radio transmitter to the back of the bat (Figs 7.10 and 7.11) so that it can be followed using a radio-receiver with a directional antenna.

Until recently, only the larger species could be tracked, due to the size and weight of the transmitters available. There is a widely accepted rule that a radio transmitter should be no more than 5 per cent of the bat's body mass, since anything greater would represent a significant aerodynamic load and may alter the bat's behaviour. Flight speed, distance flown, foraging time and other behaviours may change in response to the greater energy demand. The transmitter should also be streamlined to minimise disturbance to the air flow around the bat in flight. Progressive miniaturisation of electronic components now makes it possible to construct transmitters of 0.25–0.5 grams and all British bats can be tracked. Despite their size, these tiny transmitters can be very effective, with detection ranges of several kilometres and a life of 1–3 weeks under ideal conditions. There is often a compromise between range and life span, since battery size is the greatest limitation. Increasing the strength of the emitted signal to increase detection range reduces battery life. In practice this is not too big a problem for several reasons. Bats frequently and frustratingly groom the transmitters off before the battery expires. Furthermore, effective range is reduced in all but the flattest terrain: in hilly terrain the theoretical maximum range is an irrelevant hope. In many studies each bat need only be tracked for about half the life of the battery. For example, radio-tracking may be used to find roosts or primary foraging sites, requiring infrequent radio contact and perhaps only daytime tracking. Other studies may involve the collection of detailed and quantifiable behavioural data, demanding all-night tracking, from emergence to return to the roost, each and every night. Unless you have the luxury of many co-workers, after a week or two of this you may secretly hope that the battery will fail or the transmitter will fall off the bat so that you can rest!

What to measure

Beyond simply watching bats, there are many straightforward questions to ask, and, with thought and effort, answer. This section outlines some of the best approaches and protocols. Some of these are worthwhile projects in their own right, but the next section shows how they can be combined to build major, long-term studies. Whatever their size and scope, such projects can give meaning and structure to your interest in bats and are a good way of learning more about them, prompting better understanding and greater curiosity. There is more satisfaction in seeing something for yourself than in reading it in a book. If done well such projects can make significant contributions to our understanding of bats. Long-term studies can be particularly valuable, since the resources available to professional biologists for such work continue to decline. In the highly competitive world of scientific research, funding goes primarily to projects with short-term goals and increasingly into projects perceived to have economic or human benefits. I will take the risk of explaining the thinking behind many of these projects and protocols in some detail and trust that this will be of interest. My hope is that it will help to demystify the science and encourage those with an interest in research (but limited experience) to get involved in more quantitative natural history.

Roosts

In many respects, studying roosts, particularly those in buildings, is an easy place to start and it can provide information of direct relevance to practical

Fig. 7.12 Bats emerging from roosts in a building.

conservation. The tasks are often simple, you can concentrate on one thing at a time and you can often work in comfort. Some questions require only that you sit where you can see the entrance and count. Early emerging species, such as pipistrelles, can be seen easily, particularly when silhouetted against a dusk sky. In conjunction with a bat detector, counting is straightforward. Late-emerging bats can be harder to count, particularly if the roost exit is in the shade of a tree, and a bat detector is usually essential. Bats heard, but not seen emerging, may come from other roosts or may have taken a quick spin around the roost. However, these are likely to be in a small minority: most bats leaving the roost waste no time in flying to their foraging sites.

Where do you start? You might start by simply counting the number of bats leaving a roost at regular intervals to see whether there are any obvious patterns, before trying to find out why such patterns exist. When is a roost used, by how many bats, how does the number of bats fluctuate with the weather, over a season, or from year to year? However, this could prove to be frustrating and wasteful of time. The best science usually builds on what is already known by posing biologically interesting questions, that you then try to answer: the creation of an hypothesis that is then tested. The aim is to collect your data in such a way that you can test your hypothesis effectively with minimum effort. The best way to illustrate this is with a few examples. These examples will initially be very simple and if you are already familiar with basic bat behaviour, they may seem a little naïve. However, it is best to start with simple hypotheses – it is surprising how many of them lead to interesting observations. Only if your simplest hypothesis proves to be completely unsupported by your data do you refine it or throw it away completely. I will try to illustrate this process of refinement too. Although your hypothesis, and your procedures, may be simple, they need careful thought. Thinking about potential pitfalls, and designing your study to circumvent them, is vital. I will use the first example to illustrate the degree of complexity inherent in even apparently simple questions.

Hypothesis 1: *the number of bats occupying a roost varies over the course of a summer.*

There are several reasons why you may be interested in testing this hypothesis. You may have a straightforward, practical reason for establishing the pattern: perhaps to determine when it is best to count the bats each year to look at long-term population trends. Alternatively, establishing a pattern may be just the first step in trying to explain why it exists. Unravelling the biology underlying the pattern, in terms of the bats' life history cycle, will give you a better understanding of your local bats. For example, you might predict a gradual increase in roost size during the spring as pregnant bats gather to give birth with the benefits to be gained from living in large colonies. This might be followed by a rapid increase in the number of emerging bats as the young begin flying. You might even predict that the number of bats will approximately double, since most females will produce a single pup. In late summer you may expect to see a decline in the size of the roost as the bats disperse before hibernation.

Before beginning to collect your data, you have to plan your study. This is best done by asking questions. How often should I do a roost count to establish the pattern? Every night is too demanding and a potential waste of time. Once a month may not be often enough. Once a week or even once every two weeks would be adequate in most instances. How accurate do I need to be? There is no point in worrying about every last bat in a roost of hundreds: small errors will not alter the broad picture. What other factors may have to be taken into account? Weather is the most obvious: wind, rain, cloud cover and temperature may have an important effect on emergence, so it is worth taking notes on these. You could choose to count only on warm, still nights. If you can measure temperature and light, do so: the small, extra effort may repay you handsomely. Note down details of the roost: compass direction, structure, height above ground, protection from potential predators. How might I make the most of the data? Rather than simply counting all the emerging bats, why not time each bat out from a small roost, or count the number leaving in two or five minute intervals for larger roosts. This requires little more effort, but can be much more informative, as we will see below.

Scientists are often not able to prove their hypotheses: they merely hope to show that their data are consistent with them. For example, you may predict that as the young start flying in July, there will be a sudden increase in the number of bats emerging from a roost. If you see such an increase, you have not proven your hypothesis, but your data do support it. To test it further you must catch emerging bats during this period to see whether the increase is coincident with the appearance of young bats. Catching a few bats will strengthen your hypothesis: catching a significant proportion will be better still. However, you have to do this with appropriate care. Too much disturbance may at the very least disturb the very pattern you are trying to establish and spoil your study, at worst you may drive the bats away. Progress in science almost invariably occurs in small steps, as hypotheses are refined and tested and occasionally abandoned for new ones. Most questions, whether successfully answered or not, lead to new questions.

Assume that the seasonal trends are as you predicted, but that there are some odd perturbations. Are these perturbations random, the result of factors too complex to resolve, or can they be explained on the basis of simple environmental factors, such as the weather? If you have collected the right information you may be able to look for correlations – for example, are sudden changes in the number of emerging bats coincident with significant changes in weather? One such occurrence may be no more than coincidence, but repeated occurrences may repay closer examination. It is beyond the scope of this book to go much deeper into scientific method, but I have suggested some further reading for those who wish to pursue it further. Being able to represent your results graphically is an essential step towards understanding what your investigations are telling you. They also enable you to convey effectively your findings to others. The next step may be to apply some basic statistical tests. This need not be a daunting task with the aid of user-friendly computer software: the most difficult step is knowing what tests to apply when.

I said that we would start with simple, even naïve hypotheses and investigations. Complications to my first simple investigation may already have occurred to you. For example, a colony of bats may use more than one roost at the same time and numbers may fluctuate at a given roost as the colony fragments and circulates around its various roosts. You may throw up your hands in despair in the face of such complications: how can I possibly determine the size of this colony in the face of such problems? Alternatively, you can see them as a challenge. If you know that bats of a particular species use several roost sites in your neighbourhood you could organise simultaneous counts at these roosts. Does the total number of bats at all roosts remain roughly constant? Is a drop in the number of bats at one roost matched by an increase at another? If so, you might have evidence to suggest that all these bats belong to the same colony. Such a study can yield information of real conservation value. How would you go about testing this idea further?

This section on roost studies will be completed by briefly posing some of the many questions that could be asked.

Hypothesis 2: *the emergence time of bats is determined primarily by the timing of sunset.* The most important practical point to raise here is how do you define emergence time? Many early studies were based on the time of emergence of the first bat, but this is a poor indicator of behaviour, since it uses one potentially renegade individual to describe the behaviour of all the bats in the roost. A better approach is to use a measurement that is more truly representative. There are several possibilities, but one that is relatively easy to determine and informative is the time when half the bats in the roost have emerged. Counting the number of bats emerging in five minute intervals will allow you to calculate this. Chapter 4 described how emergence time appears to be a compromise between emerging early, when insects are abundant, to maximise foraging success, and emerging late when diurnal predators are fewer, to minimise predation risk. Given this, you can broaden the scope of your investigation. For example, do factors that alter ambient light levels, such as cloud cover, vegetation and roost orientation affect emergence time? Does emergence time vary as the number of bats emerging from the roost changes? The risk of predation to an individual bat declines as the number of emerging bats increases, simply through safety in numbers. Large roosts may therefore be expected to emerge earlier. If large roosts take longer to emerge in any case, through sheer pressure of numbers at the roost exit, then early emergence may be favoured without a change in predation pressure. How could you distinguish between the two possibilities?

Before moving on to a different type of question, here is a real puzzle about returning bats for you to consider. Why do bats swarm around the roost entrance at dawn instead of going straight in? As mentioned previously, this behaviour increases the risk of predation and wastes energy, but many bats do it, and it almost certainly has a function. However, I have not heard a plausible explanation and to my knowledge, none has been tested.

Most of the questions raised so far can be investigated at a single roost. Studying more roosts increases the possibilities enormously. Where do bats roost and do different species show different preferences? What are the most common attributes of roosts used by a species? Getting rough answers to such questions can be relatively straightforward and worthwhile. Scientific and statistical rigour is more difficult. There are many factors that could be taken into consideration. A few of them are listed here, without reference to whether or not they have been studied in the past. None have been explored fully and misconceptions and prejudices are rife in this area: there is much still to be learned.

The size, orientation, elevation and structure of the roost entrance. Do the bats prefer a secure entrance they must squeeze through, or do they prefer to fly in? Is height above ground important? Do they need a horizontal or vertical surface to land on? Are entrances more or less likely to face the setting sun? Are they clear of obstructions to give easy access or protected from aerial predation by overhanging trees?

The size and position of the roost itself. Orientation to the sun or position in relation to artificial sources of heat may be important. Roost temperature, if it can
be measured without disturbing the bats, is a worthwhile variable to study. Is it
related to the number of bats using the roost? One of the most vivid memories
of my early days visiting roosts in rural Fife is the strategic positioning of roosts
close to central heating systems or the fashionable Aga. In one hospital roof
pipistrelles roosted on the insulation around a large central heating pipe – and
on cooler days they often crawled in between the pipe and the insulation. The
age of a building often determines its structure: long-eared bats are more often
found in old buildings with large roof spaces. This may be because they prefer
to be able to fly in the roost, often changing roosting position.

Roost locality and surrounding habitat. A good roost must also be near to good
foraging habitat. But how close is close and what is good habitat? Given the
variation in flight speed, flight efficiency, foraging style and food preferences,
different answers might be expected for different species. Informed guesswork
about what habitats are important may prejudice your investigations, so try to
be as objective as possible when planning your study. If, for example, you
assume bats do not feed along major roads and ignore them, then you will
never know whether your assumption was correct. One study has shown that in
Kent Leisler's bats forage extensively along tree-lined major roads. Start by cat-
egorising all habitat around roosts, using simple categories: semi-natural woodland, plantation, mature hedgerow, pasture, arable, urban, suburban, riparian,
etc. Ask the question what habitats are present within say 0.5, 1, 2 and perhaps
5 kilometres of each roost, and how much is there of each. You will need some
simple measures of abundance for these. Again, do not make too many
assumptions. Although most studies show that semi-natural woodland is important to bats, is total area or perimeter length the most important indicator?
Bats are known to forage along boundaries, so perimeter may be even more
important than area. A few careful hours with Ordnance Survey maps will need
to be supplemented with a few more looking at the landscape itself. Having
collected your data, the secret to success is displaying them in a form that
reveals the hidden patterns. How you construct your tables and graphs will be
determined by what you are looking for, but be inventive: explore all the
options that make biological sense in your search for patterns.

Finally, much of the hard work can be taken out of roost emergence monitoring by using automatic loggers. You need an inexpensive bat detector kit or
an existing heterodyne bat detector, a home-made or off-the-shelf 'bat
counter' and an off-the-shelf count logger. The bat detector picks up a bat pass
(p. 182) and sends the audio signal to the bat counter, where it is converted
into a discrete pulse that is passed on to the count logger. This can then log
each bat pass for days or even weeks, before it needs to be downloaded and
reset. Logging time is limited largely by battery power to the detector, since
loggers that will fit into the palm of your hand are capable of storing 16,000
events over as long as three years, and a commercial bat counter will run for
up to three months. Automatic systems work well for emergence when bats
almost invariably leave the roost and fly away. Returning bats swarm around the
roost entrance before going in, so your logger will give you a largely meaning-
less mass of data, but it may tell you when and for how long swarming occurs.

Bat boxes and other artificial roosts

There has been an enormous investment in artificial bat roosts, particularly small bat boxes attached to trees (Fig. 6.3). However, relatively few serious attempts have been made to assess their value, or the relative value of different designs. It would not be worthwhile studying small bat box schemes, but large schemes are well worth investigation. The minimum viable size for study will depend upon occupancy rate, but any scheme of more than 50 boxes will repay careful planning and documentation. Smaller schemes may be suitable, particularly if compared to similar schemes run by other groups. Think carefully before starting. If you use a single box design, place boxes in such as way as to be able to compare occupancy in relation to orientation relative to the sun, height above ground, tree species, entrance obstruction by vegetation, microhabitat (for example, woodland edge versus interior) habitat type, and so on. But do not try to be too ambitious. Unless you have hundreds of boxes you cannot investigate more than two or three of these variables at a time. You will need at least ten boxes in each category, perhaps more, and each category must differ in only one respect from other categories. In other words, you cannot compare woodland edge boxes 10 metres above the ground with woodland interior boxes 3 metres above the ground. However, if you *also* have 3-metre high woodland edge boxes and 10-metre high interior boxes, you can study both factors at the same time, but already you need a minimum of 40 boxes. Having designed your layout, you need to check your boxes frequently, regularly and perhaps over a long period. By inspecting every month from say April to October you can not only collect a reliable database to investigate box siting variables, but also to look at how quickly the boxes are occupied after siting, and seasonal variations in occupancy rate.

Next comes the thorny question of how you inspect the boxes. The more information you collect the more valuable your data. Number of bats, species, sex and individual identity are all potentially interesting and valuable, but all involve disturbing the bats. Too much disturbance may influence roosting patterns and even drive bats away. A brief inspection with torch and mirror, or a gentle lifting of the box lid, will usually provide number and species with negligible disturbance. Anything else clearly comes under the category of significant disturbance. My own experience working with the Forestry Commission suggests that careful handling of the bats does not alter their behaviour very much and certainly does not cause them to abandon the boxes. The number of bats using the boxes remains steady over many years and ringed individuals turn up time and time again in their preferred boxes. However, many ringed bats are rarely seen and it is very difficult to determine whether this is due to natural dispersal, mortality or our disturbance. It would, of course, be possible to study this directly by setting up and monitoring two identical schemes, in which bats are left largely undisturbed in one and fully identified and sexed in the other.

Regular, detailed inspection can also tell you why bats use the boxes, as nursery or mating roosts for example. To avoid undue disturbance, many people do not inspect their boxes during the nursery season. If you think it is important to establish what types of boxes are used as nursery roosts, a brief inspection when the young are grown, to confirm your suspicions, may be an acceptable compromise.

It has sometimes been suggested that bat boxes may actually draw bats into a previously unoccupied habitat. The only way to test this hypothesis is to quantitatively measure bat activity *before* and *after* introducing bat boxes to a site. This has yet to be done.

Measuring bat activity

Habitat use: foraging and commuting

Before the appearance of bat detectors the most common method of estimating the number of bats using a particular site or habitat type was to catch them using mist nets and harp traps (p. 171). This is certainly the most reliable way of identifying species, but it does have serious drawbacks. Capture success is very dependent upon the skill and experience of the biologist in choosing netting sites and erecting nets and traps. Even the best placed nets will catch only a small proportion of the foraging and commuting bats in the area. Catching can also be very selective. Nets and traps are most conveniently placed within 5 metres of the ground where they catch only low-flying bats, although (with some difficulty) they can be set higher. The catching area is small: a single mist net is only about 3 metres high and rarely longer than 30 metres, and harp traps are rarely bigger than 3 metres by 2.5 metres. Differences in foraging style and flight/echolocation performance can make some species easier to catch than others. Quantifying relative bat activity or species composition, in terms of bats caught per hour per metre of net, is theoretically possible, but subject to many potential pitfalls. It is also very hard work.

Quantifying bat activity with a bat detector is much quicker and easier. A bat detector, in good atmospheric conditions, will detect bats in a broad arc at distances of up to 30 metres or more. Wind, rain or dense vegetation will reduce the range. A heterodyne detector (p. 161) will be selective, but two such detectors, tuned to say 45 and 25 kHz, will pick up all but horseshoe bats. A frequency division or time-expansion detector (p. 164) will in theory detect all species. However, species with high intensity calls such as noctules will be detected from further away than most, and long-eared bats may not be heard at all.

We start with a simple hypothesis: *bats (all species) feed in some habitats more than others*. To test this you need a quantitative measure of activity and an experimental design that samples all available habitats with equal intensity. What is our best measure of activity? Since foraging is the most important activity to measure, in that it is generally the single most important reason for the presence of a flying bat, counting feeding buzzes is a good start. If the number of flying bats is low, too few feeding buzzes may be heard to give reliable results. Since a flying bat is more likely to be feeding than doing anything else, you could simply count bat passes and assume the bat is looking for food. You can test the validity of this argument by seeing whether the number of feeding buzzes increases in proportion to the number of passes. But what is a bat pass? This is usually defined as the discrete burst of echolocation calls heard as a bat flies into the range of your bat detector and out again. It is usually a well-defined event lasting a second or so, with a clear gap between it and the next, but you may get little more than a few chirps as a bat flits through. When bat activity is very high one pass may merge into another and defining a pass is more difficult. The exact definition of a pass in these circumstances is not too critical, but it is important to be consistent. Counting passes or buzzes does not

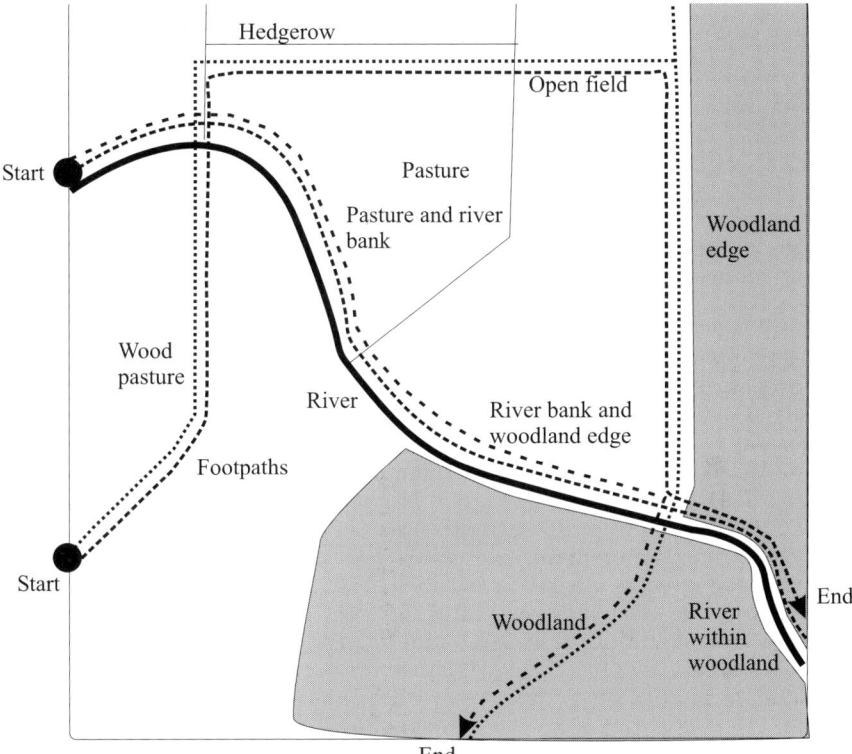

Fig. 7.13 A hypothetical bat transect through a complex habitat.

count the number of bats, since a single bat may pass you several times or several bats may pass only once. However, it does give a relative measure of foraging activity that will enable you to compare the relative importance of different habitats to feeding bats.

Since it involves little extra effort, it makes sense to count both passes and buzzes and increase your options. The next question is, how do you sample different habitats in a meaningful way? There are two approaches: counting from fixed positions or walking transects. If you count at fixed positions you need to choose your positions and the length of time you sample for. Assume you have a few simple habitat types as shown in Figure 7.13: woodland, wood pasture, pasture and river bank, with several footpaths running through the landscape. The simplicity is deceptive and this map illustrates the pitfalls.

You should sample for the same length of time in each habitat to enable your results to be compared: long enough to get a representative measure of activity, but not too long, since it is important to sample all habitats under near identical conditions. Only differences in habitat should explain any differences in activity observed. A change in temperature, the number of flying insects, or the foraging strategy of the bats through the night could complicate the interpretation of your results. Ideally, all habitats would be sampled simultaneously, but you cannot be at several places at once. You can overcome this problem by rotating the order in which the sites are sampled each night. Since reliable results will only be obtained if you sample over several nights, you will have to repeat your observations several times anyway. Working in teams can help, but you have to be sure all members of the team use detectors of similar sensitivity and follow the same protocol. To be doubly sure, you would rotate your team around your habitats.

Assume you are going to sample each of your habitats in the second 60–90 minutes after sunset. This allows time for virtually all bats to emerge and begin foraging, but limits the time available for significant changes in weather, insect availability or bat behaviour. You could start with recording from fixed positions. It would perhaps be foolish to spend all the time in each habitat in one place. Maybe you should divide your time in each habitat between two to five random locations, or locations chosen to be representative of the habitat? On the map shown, the habitats might be wood pasture, pasture, river bank, hedgerow, woodland edge and inside woodland. Do you include subdivisions such as river bounded by woodland or pasture?

Alternatively, you can take a slow, meandering walk through the landscape, recording as you go. If you choose the latter, you must spend an equal amount of time in each habitat, or at least correct for differences by expressing your results as bat passes per kilometre or per hour. With careful planning you can cover all your habitats in one or two continuous walks, marking each bat heard on a large-scale map. Using maps can enable you to pinpoint key foraging sites on a local scale, producing persuasive, quantitative evidence to show that a particular wood, pond or other feature is of particular importance to your local bats. Carefully planned and repeated from year to year you can track local changes in bat populations. They may reflect real changes in numbers or changing distributions, but if matched by significant changes in land use, they could be informative. Careful work of this nature can tell you a lot about bats and be a persuasive argument for conservation on a local scale. However, there are complications. Most routes through this landscape sample two or even three habitats at any one time. For example, a walk from east to west along the river first samples river bounded by pasture and wood pasture, then river bounded by pasture and woodland edge and finishes with river within woodland. The second east–west transect is no less complex: wood pasture, hedgerow in pasture, pasture, woodland edge in pasture and woodland. Do you simplify this? For example, the first transect could be river without trees followed by river with trees. I find it best to simply mark on copies of the map all bats heard, as they are heard, as you walk the transect. Analysis can begin with simple divisions, such as trees/no trees along a river, which are likely to prove the most informative. However, you still have the raw data for finer analysis should you feel it worthwhile attempting it. That said, you must still plan your transect carefully to meet all the conditions described above.

Depending upon the type of detector used, you can sort your bats into groups or even species, as described earlier (p. 161). There are all sorts of variations on this theme, with many habitats to work on – but keep your categories simple. You can see how important your local parks are relative to arable farms, whether rivers above and below sewage outfalls have different numbers of bats, how new buildings and new lighting affect bats, and so on.

A big factor influencing the activities of British bats is our unpredictable weather. Any change in the weather that alters the number of flying insects can potentially alter bat behaviour. In the case of all the projects suggested so far, you should pick relatively mild, still and dry nights. But you may wish to investigate the effects of weather itself on bat activity. Temperature, rainfall, wind speed and direction are easily measured or estimated. It never ceases to amaze me how bad conditions need to be before bats are driven back to roost.

The landscape is also important to commuting bats and there are ways to study this behaviour too. It is not so easy to quantify, but it is worth a brief mention. At dusk, in open landscapes, it can be relatively easy to identify commuting bats. As a general rule they have a fast, purposeful flight and will often follow hedges, lanes, tree and water lines and similar features to their foraging sites, rather than cross open fields. This strategy may reduce predation risk and may facilitate navigation. Repeated observation around the periphery of large roosts can enable you to map commuting routes.

Finally, think about safety. Be sure to plan and test your route in daylight, to minimise the likelihood of accidents. Take a friend and let someone at home know where you are going and when you plan to be back. If your late night wanderings are likely to arouse interest it is advisable to consult landowners and the police, even when you are using public footpaths. Even after careful consultation my research team is still occasionally taken for poachers!

Hibernation

Studies of hibernating bats must be carried out with extreme care, since disturbance can have catastrophic effects on the value of the hibernation sites to bats and on the bats themselves. However, there is much interesting and valuable work that can be done with little or no risk of disturbance. Perhaps the most important activity is identifying hibernation sites. Only when they are identified can they be given adequate protection. Finding hibernation sites for those species that do not go underground is like finding needles in a haystack and any efforts are unlikely to be worth the rewards. For this reason only underground hibernacula will be discussed. Before humans came on the scene, this meant caves and rock crevices, but it now covers a wealth of additional possibilities: mines, rail and canal tunnels, bunkers, basements, ice houses and a variety of industrial sites. Almost any cool, dark, damp, accessible and undisturbed structure is a potential hibernaculum. If you live in an area of the country rich in natural caves, some of the more significant sites may already have been identified. However, many others will be unknown. In other areas of the country, very few sites will have been found. Although large caves and mines are arguably the most likely sites to be used, many small sites can be equally important. The crucial factors are degree of disturbance and microclimate. Even small sites may have a suitable microclimate if the structure and air flow are right.

To find a hibernaculum you have to find the bats *in situ* or observe them going in and out. Droppings are a good clue and may be seen at any time of year. The first step in both cases is to identify possible sites, through local knowledge, by studying maps and by visiting the sites. Many will be potentially dangerous, on private land and attract other interest groups. It is very important to do your homework before beginning your studies, to avoid danger or conflict.

The safest approach (for bats and biologists) is to identify sites from the outside, by visiting them at the right time of year with a bat detector. Before going

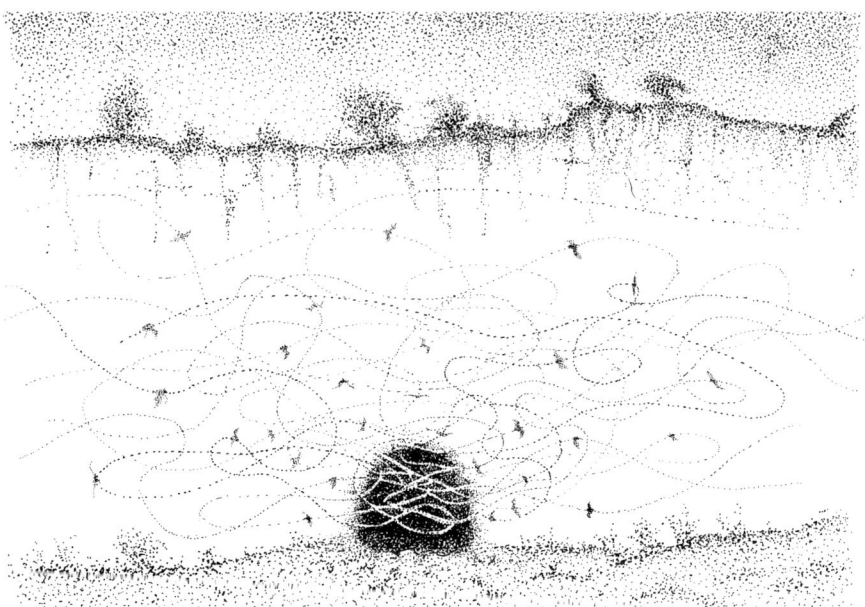

Fig. 7.14 Bats swarming around the entrance to an old mine.

into hibernation, many bats make repeated visits to hibernacula in late summer and autumn. The bat traffic around the entrance to a hibernaculum, measured in bat passes, can be very heavy – much heavier than in the surrounding habitat. Bats do not simply arrive and fly in, but swarm around the entrance (Fig. 7.14). Activity rarely peaks until several hours after dark and can be easily missed if your visit is early in the night and brief. There is also considerable seasonal variation in activity. In North Yorkshire, where I largely work, peak activity is typically from late August through September, declining steadily through October and November. The best time to find new sites is therefore September. However, there is year-to-year variation and different species appear to peak at different times. Pre-hibernation swarming sites may not always be major hibernation sites, but many are, and identifying swarming sites is in any case important in itself for reasons discussed in Chapter 4.

As with bats emerging from a summer roost, it is worthwhile recording bat passes with time, to establish when activity does in fact peak during the night. This allows more meaningful comparison between nights and sites and may give you clues about its significance. In my experience, as the season progresses, relative activity around dusk increases, suggesting more of the bats are roosting in the cave, rather than coming from more distant sites and arriving later. This is backed up by the relative numbers of bats entering and leaving the site. You can investigate this yourself if the entrance to the site is not too large and can be covered by a night scope or video camera. By recording bat passes and the number of bats flying into and out of the site over the course of a night you can collect a lot of information with no disturbance. If the site is used for hibernation, the net flow into the site should increase as winter approaches. Bats emerging from hibernation in the spring rarely linger, but it is possible to do a follow-up in, for example, April.

Most of the known hibernation sites in Britain appear to be home to a relatively small number of bats: we have few sites comparable to those found in continental Europe or North America, that can house thousands, tens of thousands or even hundreds of thousands of individual bats. Bear in mind, however, that we may be underestimating populations of crevice-roosting species. Whether large or small, it is important to identify these sites, particularly since many are vulnerable to disturbance, destruction or closure. Because each may be used by five or more species, they are also interesting to visit.

The biggest difficulty in this sort of study is finding out which sites merit further investigation without spending many nights in fruitless pursuit of bats before striking lucky. Experience will count for a lot, but this does not come quickly. One way to speed up the process is to use automatic loggers as described at the end of the section on roosts (p. 180).

Having established the importance of a site, you may wish to confirm the species present, which can only be done unambiguously by catching them. Too frequent capture at any one site should be avoided, but these sites may turn up real rarities: on a single night at a site in the Southwest we caught eight species, including Bechstein's bat, which had not previously been recorded in the area.

Individual behaviour

Some of the most fascinating studies are the more intimate: those that involve investigating the behaviour of individuals in some detail. There is enormous satisfaction and insight in radio-tracking a bat night after night, slowly building up a picture of its day-to-day routine. This inevitably means capturing and marking the bats in some way, so a licence will always be necessary, whether your project involves fur clipping, ringing, or the use of light tags or radio transmitters. Because of this, and the complexity, difficulty and potential disturbance to the bats, a discussion of these topics is beyond the scope of this book. However, if you are enthusiastic and determined, it is possible to get involved in this sort of work.

Recording your data

Whatever projects you pursue, if you have interesting and reliable findings you should let other people know about them. Natural history and science are of little or no value if they are not communicated to others. Sharing your findings also enhances the pleasure you derive from your efforts. It leads to discussions with other informed bat watchers that may increase your knowledge and understanding and give you new avenues to pursue. Your findings may interest others and prompt new work from them. Finally, the knowledge may benefit the bats themselves through increased public understanding and better conservation. So, keep good records and share your data and ideas.

Some loose ends

There are frequent local and national projects that you can become involved in and the BCT usually has information on these.

Many of the activities described will involve contact with people with little or no knowledge of bats, such as landowners, the police, local government offices, and so on. These people may either not share your enthusiasm for bats or be very busy when you contact them. Be well informed and organised before seeking help or permission to visit sites and be courteous and patient in your dealings with them. Contact your local bat group if you are new to bat work. Buy a copy of the *Bat Workers' Manual.*

Finally, do not do anything that disturbs the bats you are studying unless you are sure the information is valuable and justifies the disturbance: and the information is only valuable if collected with thought and care.

Practical conservation projects

The projects described in the rest of the chapter are also very practical, in the sense that you have to go out and do them, and most are important to bat conservation, but they all focus on learning more about bats. This section discusses what can be done to promote and conserve bats *directly*. Not everyone wants to spend long nights chasing and studying bats, but you may like to help conserve bats in other ways. Whatever activity you get involved in, it is worth joining your local bat group for the information and assistance you can get from it, and the contacts you will need if you require a bat worker licence.

Education and raising awareness

As discussed in Chapter 6, many threats to bats are underpinned by ignorance and misinformation. A major part of bat conservation is therefore making the world aware of the fact that bats are there, that they are interesting and that they need conserving. There are many ways in which this can be done and what follows are just some of the more obvious: the more extreme methods used to attract attention to bats have included bungy jumping in a bat suit!

Teaching and talking in schools

If you are a teacher, or have some teaching experience, working in schools is rewarding and effective. Children are almost universally positive about bats and keen to learn about them. They can in turn have a very positive influence on their parents, who may initially be less receptive. Talks with good slides are always welcomed by schools, particularly if you can take a few bats from your local bat carer. The Bat Conservation Trust (BCT) and Bat Conservation International (BCI) have stunning slide packs with notes on each slide. Both also produce teachers' resource packs and the BCT pack integrates bats into the National Curriculum. The BCT also has an education officer who may be able to help you.

Talks

Giving a good talk requires practise, confidence and a few skills that not everyone has: but you soon learn whether you have them or not! However, with a little natural aptitude, good organisation and above all *enthusiasm*, many people can give effective talks. The task is made a lot easier if you have good resources, in particular slides, but also sound and video and perhaps some practical demonstrations and props. In addition to their slide packs, BCT and BCI have a wide range of resources, including booklets, leaflets, posters and stickers. They also have bat-related items that can be sold to raise funds for conservation work. There are simple things you can do to illustrate some of the more unfamiliar concepts in bat natural history. To show just how intense bat echolocation calls are, I take a domestic smoke alarm to talks: most emit sound at the same intensity as bat calls. A child's slinky spring can be used to show how sound waves travel: when stretched horizontally, a short, sharp push on one end will send down a visible pulse and return an echo. I can demonstrate echolocation with an inexpensive ultrasound 'tape measure' (as used by estate agents) strapped to a x32 time-expansion bat detector. The ultrasound pulses are clearly audible to the audience and the pulse-echo delay is increased to such an extent by the x32 expansion that the pulse and its echo are easily distinguished. Using this device, with a little practice, you can walk blindfold around a room and avoid the walls.

Walks

If you are looking for a big return on your investment, you cannot do better than an evening emergence watch or bat walk or both. It is an excellent way of raising public awareness without having to put yourself in the spotlight. Success depends on the reliability of your performers: many bats and these turning up on cue. You therefore need to reconnoitre your sites. For an emergence watch, a good strategy is to make arrangements with a proud 'roost-owner': someone with a large colony in their home who is happy to show them off to a group of people. You settle down comfortably just before dusk, in a good position to see the emerging bats, and wait for the performance to begin. You should ideally have enough bat detectors to let everyone present try one out. A night scope or night-vision camcorder on a tripod, focused on the roost exit, can give you a close-up view of the bats squeezing out and leaping into flight. Bat walks are best done along a stretch of still, narrow river or canal, or along the bank of a small lake, to maximise the number and variety of bats. The more people attending, the more 'guides' and bat detectors you need – you can never have too many. A powerful hand torch, used with discretion, is very good at spotlighting bats feeding over water. One memorable day some years ago we combined a talk in a village hall with a barbecue and emergence watch, followed by a bat walk along a river – a real success.

Bat box days

If you run a bat box scheme, you can run bat box open days – providing you can guarantee your visitors some bats in a pleasant environment. Simply publicise one of your regular inspection days with instructions on where to meet, an idea of how long the event will last and what your visitors need to bring by way of clothing and food. This is an ideal way of showing bats close up with little more disturbance than they would suffer from a normal bat box inspection day.

   With all outdoor events, take along leaflets and website addresses for those who want to know more. If they have the information when they get home they will be enthusiastic enough to follow it up straight away. Send them the information a week later and for some the initial enthusiasm may have cooled.

Hands-on conservation

Bat hospitals

Many people have particular skills that they can use in bat conservation. Scattered throughout the country are bat carers and bat hospitals, looking after injured or orphaned bats. Rehabilitation and release of such bats is unlikely ever to have an effect on natural populations, but its value to bat conservation lies in other directions. The skills acquired with common species may be invaluable in the future care of rarer species: perhaps even in captive breeding programmes. A roost full of orphaned Bechstein's bats represents a significant proportion of the known UK population and rehabilitation to the wild would be a worthwhile venture. As a public relations exercise, bat care is a big seller and features regularly in the media, invariably casting bats in a positive light. Without bat carers we would not have a supply of bats with which to enchant our audiences when we give talks. You do not need a licence to look after injured bats, but you must keep detailed records to show where your bats came from. Ideally, a fully recovered bat should be released into the wild. It is assumed that an injured bat that has been in captivity for more than two to three weeks will be unable to fend for itself in the wild. Although difficult to test, this rule should be tested, if rehabilitation is to be a real goal. Many of the groups working in bat care share information.

Construction

If you have woodworking or other construction skills, there are projects to challenge your abilities, from building simple bat boxes through customised larger bat boxes to roost modification and even building from scratch. If you have these skills there will be someone out there waiting to make use of them. Bat box designs are available from both the BCT and BCI. Knowing what bats need from a roost means that you can try your own designs out. The standard bat box design that has dominated the scene for 30 years is probably not ideal, with its large internal cavity and lack of snug crevices. Grille design and construction for hibernation sites, remodelling of roof spaces and the construction of roosts for the outside walls of buildings are in increasing demand. Some ambitious projects need the services of architects as well as builders.

Electronics and computing

As this chapter has shown, bat study is becoming increasingly technical and technological. Many of the tools needed are expensive, or cannot be bought off the shelf. If you can put together bat detector kits or set up bat and environmental loggers you have a valuable skill. Keeping records, setting up databases and analysing data are computer skills some bat groups still lack.

Organisational and fundraising skills

A valuable part of any conservation machine is the organising of group activities, raising money and managing funds. If you are good at these activities, there is a bat group that needs you.

Habitat restoration and creation

Finally, there is the creation of new habitats and the restoration of degraded sites. This involves primarily hard physical work such as tree planting and pond digging. The National Trust, RSPB, British Trust for Conservation Volunteers and others all organise this kind of work.

Appendix

Identification: How to Identify a Bat in the Hand

Anyone who becomes seriously interested in bats will soon want to be able to identify them in the hand. If you have never done it before, then it is best to work from a key. In time, you learn to take short cuts, but if you are ever in doubt, you should go back to a key and use all the features available to you: do not rely on a single indicator with difficult species. If you are teaching others, it is also best to start with a key. In short, keys are invaluable, so here is a simple key, based on the features I find most useful in the field.

Imagine that you are somewhere in the south-west of England, and lucky enough to be in an area that has all 16 British bats. You are taking bats out of a harp trap for inspection and identification, never having done this before. Where do you start?

1. Does the bat have a conspicuous and complex structure on its face: a noseleaf (Figs A and B)?

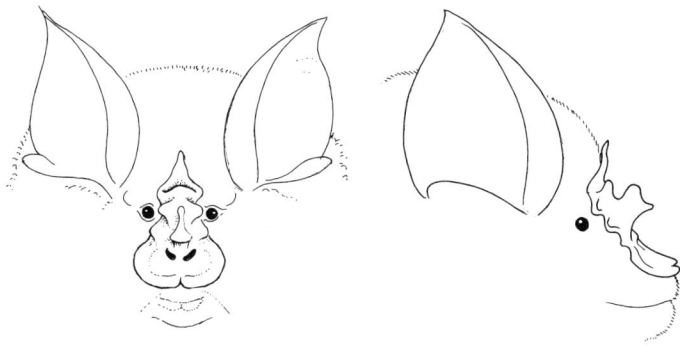

Figs A and B. Front and side views of horseshoe bat to show noseleaf.

   Noseleaf present, go to 2.

   Noseleaf absent, go to 3.

2. Forearm length greater than 45 millimetres: greater horseshoe bat, *Rhinolophus ferrumequinum.*

   Forearm length less than 45 millimetres: lesser horseshoe bat, *Rhinolophus hipposideros.*

3. No noseleaf: the bat is a vesper bat. (Figures C and D show typical faces). Are the ears joined over the head?

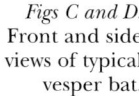

Figs C and D.
Front and side
views of typical
vesper bat.

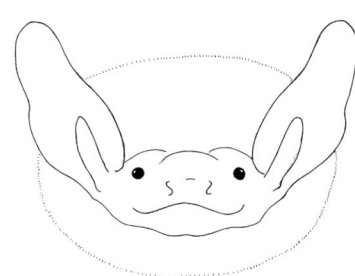

Ears joined over the head, go to 4.

Ears not joined over head, go to 6.

4. Ear less than 20 millimetres long: barbastelle, *Barbastella barbastellus* (Fig. E). This is a medium-sized bat, large almost square ears, joined in the middle of the forehead, and the eyes are almost inside the pinna. It has dark, almost black fur, usually tipped cream or yellow.

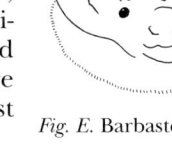

Fig. E. Barbastelle bat

Ear more than 25 millimetres long: long-eared bat (Fig. F). There are two species, the brown and the grey. Colour is not a good guide. Go to 5.

5. Tragus is less than 5.5 millimetres wide and the thumb more than 6 millimetres long: brown long-eared bat, *Plecotus auritus*.

Tragus is more than 5.5 millimetres wide and the thumb less than 6 millimetres long: grey long-eared bat, *Plecotus austriacus*.

Fig. F. Long-eared bat.

6. Ears are not joined in the middle (Figs C and D). There is a (sometimes hard to find) cartilaginous projection from the foot, towards the tail, helping to support the trailing edge of the tail membrane: the calcar (Fig. G). There may be a rounded flap of skin attached to the outside edge of the calcar; the post-calcarial lobe. Is there a post-calcarial lobe (Fig. G)?

Post-calcarial lobe present, go to 7.

Post-calcarial lobe absent, go to 8.

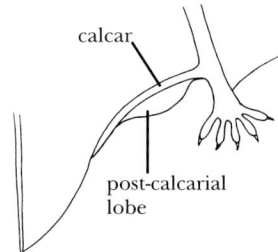

Fig. G. Calcar and post-calcarial lobe.

7. Forearm less than 37 millimetres: pipistrelle. Until recently we would have confidently assigned this bat to a single species, the common pipistrelle, with an outside chance that it was a rare migrant. We now have to distinguish between three species that are not easily separated without practice or analysis of their echolocation calls.

The '45' pipistrelle, *Pipistrellus pipistrellus* (Fig. H) typically has dark brown fur, a long black nose and a dark face. Its distinctive echolocation call (with peak energy at 45 kHz) is the best guide to identification. The fur on the penis is dull white or grey. In parts of Europe, another useful feature has been defined, but requires confirmation in Britain. The ratio of the second and third fingerbones, that is the last two, on finger three (which reaches to the tip of the wing) is 2:3.

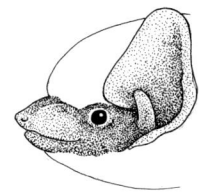

Fig. H. 45 pipistrelle.

The '55' pipistrelle, *Pipistrellus pygmaeus* (Fig. I) is typically lighter, with a flatter nose and the males have a noticeable yellow tint to the fur on their penis. Its echolocation call has maximum energy at 55 kHz. The second:third finger bone ratio of finger three is 1:1.

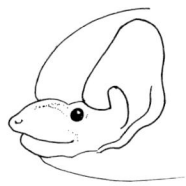

Fig. I. 55 pipistrelle.

Nathusius' pipistrelle, *Pipistrellus nathusii*. Look for the additional collagen band on the wing (Fig. J). Other identifying features are a ratio of forearm length to the length of the fifth digit of more than 1.25. Nathusius' pipistrelle typically has dark dorsal fur with light tips.

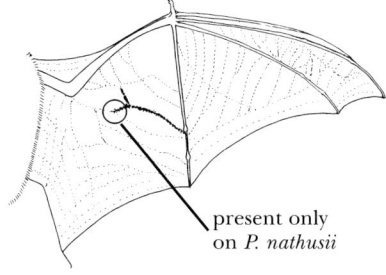

present only on *P. nathusii*

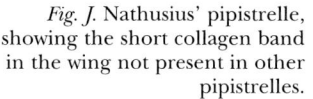

Fig. J. Nathusius' pipistrelle, showing the short collagen band in the wing not present in other pipistrelles.

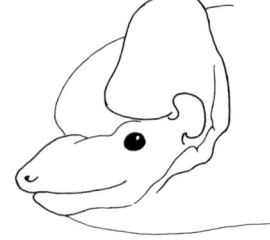

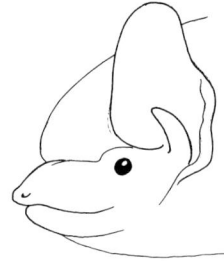

Fig. K Leisler's bat.   *Fig. L* Noctule.   *Fig. M* Serotine.

Forearm length greater than 37 millimetres but less than 47 millimetres: Leisler's bat, *Nyctalus leisleri* (Fig. K). Other features include shaggy fur with a dark base and a pale tip. It has a mushroom shaped tragus, but so does the next bat.

Forearm length greater than 47 millimetres. Tragus mushroom shaped: noctule, *Nyctalus noctula* (Fig. L). This is a large, sleek and often golden bat. About 2 millimetres of tail projects beyond its tail membrane.

Forearm length greater than 47 millimetres. Tragus shaped like a blunt banana: serotine, *Eptesicus serotinus* (Fig. M). Look also for a tail that projects 5–7 millimetres beyond a tail membrane with an s-shaped edge (Fig. N).

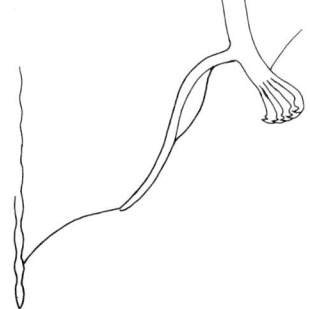

Fig. N. Projecting tail and s-shaped calcar of the serotine.

8. No post-calcarial lobe. The bat belongs to the genus *Myotis.*

Forearm length greater than 50 millimetres, go to 9.

Forearm length less than 50 millimetres, go to 10.

9. Mouse-eared bat, *Myotis myotis.* Now extinct in Britain.

10. Ear length differences now identify two further species, leaving a final group (11).

Ear greater than 18 millimetres long, Bechstein's bat, *Myotis bechsteinii.*

Ear 14–17 millimetres in length. Natterer's bat, *Myotis nattereri.* Ear long enough to project beyond nose, with a thin tragus greater than 50% of

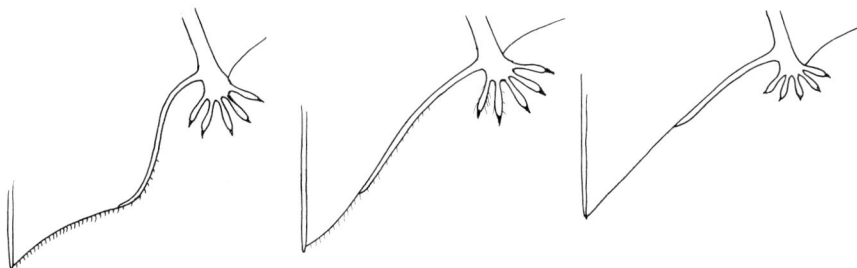

Fig. O. S-shaped calcar and stiff tail bristles of Natterer's bat.

Fig. P. Long calcar of Daubenton's bat.

Fig. Q. Short calcar of whiskered/Brandt's bats.

ear length. There is a characteristic upturn of the ear tip, an S-shaped calcar and small but **distinctly stiff** bristles along the trailing edge of the tail membrane (Fig. O).

11. Calcar length separates one of the remaining three species from the last two.

Calcar extends more than half the distance to the tail. Daubenton's bat, *Myotis daubentonii* (Fig. P). Other features include large feet, a blunt tragus and fine, soft hairs along the trailing edge of the tail membrane. The latter are sometimes hard to spot. The wing membrane may attach higher on the ankle than the next two species.

Calcar extend less than half way to the tail (Fig. Q). The bat is either a whiskered bat, *Myotis mystacinus* or Brandt's bat *M. brandtii.* These can be very hard to separate.

Outer edge of the tragus convex. Brandt's bat, *Myotis brandtii* (Fig. R). The penis is bulbous (Fig. S). With a hand lens, examine the teeth: the protocone of p⁴ in the upper jaw is taller than p³, the small tooth in front of it (Fig. T).

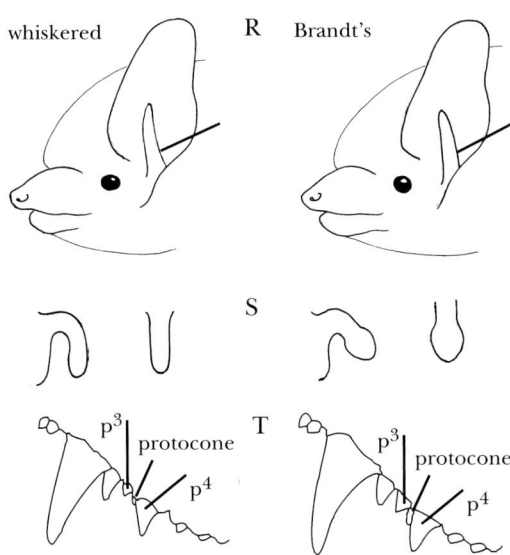

Figs R, S, T. Tragus, penis and tooth differences in whiskered and Brandt's bats.

Outer edge of the tragus concave, making the tragus pointed (Fig. R). Whiskered bat, *Myotis mystacinus*. Penis thin and straight (Fig. S). The protocone of p^4 in the upper jaw is lower than p^3 (Fig. T).

You may need to look at quite a few bats before the difference in tragus shape between these two species becomes apparent. Colour is variable, but whiskered are often dark, almost black, Brandt's a mid brown. Don't be surprised to find bats that don't fall convincingly into either camp: whiskered bats and their relative may hold some surprises. Brandt's itself was only recently separated from whiskered, and a new *Myotis* species, *M. alcathoe*, closely related to both, has recently been identified in southern Europe. I doubt it will be the last.

Glossary

Acoustic fovea: A disproportionately large part of the hearing system of a bat given over to a particular, small range of **frequencies**, making the bat acutely sensitive to sounds in this range. It is usually centred around the frequencies used by the bat in its **echolocation** call.

Adaptive radiation: The process by which organisms evolve from a small number of species into a larger number, to fill the available **ecological niches**. Such radiations frequently occur after major environmental changes create new niches.

Aerodynamic efficiency: An aerodynamically efficient wing is one that wastes little energy in generating **thrust** and **lift**.

Aerofoil: A shape that when moving through the air can generate **lift** and **thrust** with little **drag**. In cross section, an aerofoil is typically round at the front, tapering to a downward curving point at the rear.

Altruism: An act by an animal that benefits the recipient of the act, at some personal cost to itself. Altruism should not be seen in nature in its pure form, since each animal should act selfishly to ensure its genes stand the best possible chance of survival to the next generation. It is seen between related animals, because they share genes and sometimes in reciprocal form: you scratch my back and I'll scratch yours.

Amplitude modulation: Variation in the amplitude, loudness or intensity of a sound with time.

Angle of attack: The angle made between an **aerofoil** and the direction of the airflow.

Anoxic: In the absence of oxygen

Anticoagulant: A chemical which prevents the clotting of blood. Used clinically, amongst other reasons, to prevent the formation of internal blood clots that can lead to heart attacks and strokes.

Arms race: The process by which predator and prey evolve in an attempt to stay one step ahead in the race for survival. The argument to explain the evolution of sprint running in the cheetah and the gazelles it chases.

Aspect ratio: A term used to describe the shape of a wing. Wingspan divided by average width, or wingspan squared divided by wing area.

Attenuation: The loss of energy as a sound wave travels through air. Energy losses are greater at higher **frequencies**.

Bandwidth: The range of frequencies within a burst of sound such as an **echolocation** call. A **broadband** call has a wide range of frequencies.

Bat box: A small, artificial roost attached to a tree to simulate a tree cavity.

Bat detector: A device that makes inaudible **ultrasound echolocation** calls audible to the human ear.

Biodiversity: A term used to describe the enormous variety of life on the Earth.

Biped: An animal that moves on two legs.

Breakaway: The separation of the airflow from an **aerofoil** to produce **turbulence** and loss of **lift**.

Broadband: A sound made up of a wide range of **frequencies**.

Calcar: The long, thin projection of **cartilage** from the foot of a bat towards its tail, along the trailing edge of the tail membrane.

Cardiovascular system: The system that moves blood around the body: the heart and blood vessels.

Cartilage: A flexible but strong structural material produced by the body, typically found in joints.

CF: Abbreviation for constant **frequency**, to describe on of the major types of **echolocation** call, one in which most of the call is at a single frequency.

Circadian: Following a daily rhythm.

Circannual: Following an annual rhythm.

Climate change: The complex changes induced in global climate patterns by an increase in average global temperature. It is widely accepted that the increase in temperature is due at least in part to atmospheric changes caused by human industrial activity.

Clutter: A term used to describe three-dimensional complexity in the air space of a flying, **echolocating** bat.

Cochlea: The structure in the inner ear of all mammals responsible for detecting and differentiating between sound of different **frequencies**.

Coevolution: Two or more species evolving together, usually to the mutual benefit of all. For example the evolution of flowers to attract specific bats may be accompanied by specific adaptations in the bats to facilitate the collection of nectar and pollen. Both the plant and the bat may benefit from these adaptations.

Collagen: A protein present in tendon, muscle and other tissue that increases its tensile strength and resists stretching.

Commuting: Travel between roost site and foraging site, with little or no feeding along the way.

Competitive exclusion: The process by which one organism out competes another for essential resources (e.g. food and shelter), to the extent that the loser is excluded from the range of the winner.

Crepuscular: Active at dusk and dawn

CROW Act: Countryside and Rights of Ways Act, 2000. Important new legislation that should have a significant impact for good on nature conservation.

Diurnal: Active by day. Operating on a daily cycle.

Doppler shift: A shift in the **frequency** of sound (or electromagnetic radiation such as light), due to the movement of the object emitting the sound relative to the observer.

Drag: Force resisting the movement of a bat through the air due to friction and **inertia**.

Duty cycle: The proportion of time spent, for example, emitting **echolocation** calls. Long duty cycle bats emit long duration calls with relatively short gaps between each call.

Echolocation: Orientation and navigation in the environment carried out by interpretation of the echoes of specially emitted pulses of sound.

Ecological niche: A distinctive and sometimes unique way of making a living in the environment that minimises competition for resources with other species.

Ectoparasites: Organisms, typically invertebrates that make their living on the body surface of other organisms, frequently at a cost to the host, e.g. ticks, mites and fleas.

Elastin: A protein that helps make tissues strong but elastic.

Endemic: A species or subspecies found only in one locality.

Endoparasites: Organisms that make their living within the tissues of other species, usually at a significant cost to the host.

Endothermy: The maintenance of a body temperature above the temperature of the environment by means of **metabolic heat** production.

Eocene: Geological period from about 55 to 38 million years ago.

Eutrophication: Excessive plant growth in freshwater due to nutrient pollution such as phosphate and nitrate run-off from farms.

Evolution: The change and diversification of life on Earth by the processes of random mutation and natural selection.

Feeding buzz: The rapid increase in the rate at which **echolocation** pulses are emitted by a bat as it approaches and captures its prey. Heard as a distinctive buzz on a **heterodyne** bat detector.

Fitness (evolutionary): A measure of an organism's immediate ability to thrive and pass on its genes to the next generation.

Flycatching: Hunting insects by means of short flying sorties from a convenient perch.

FM: Abbreviation for frequency modulated, to describe a major type of **echolocation** call, one in which frequency changes as the call is being emitted.

Frequency (of sound): The number of air vibrations each second (in Hertz or Hz) of a sound.

Frequency modulation: A change or *modulation* in sound **frequency**. Can be caused by the bat itself in emitting the sound, or in the echo by movement of the target.

Frugivore: An animal that feeds exclusively or primarily on the fruit and related products of plants.

Fundamental frequency: The lowest natural **frequency** component of a sound.

Gestation: The period of growth of the foetus in the womb, from fertilisation to birth.

Gleaning: The capture of stationary prey from the ground, leaves, or other surfaces, during which the bat may hover or even land.

Ground effect: Increased **lift** achieved by flying very close to the ground or to a water surface.

Group foraging: Foraging in groups may increase foraging success by making it easier to find patchy food sources or by increasing the rate of capture of elusive prey.

Habitat fragmentation: The break up of large patches of a particular habitat into small, isolated fragments that are often too small to sustain the ecosystem that was stable in the original large patch.

Haemoglobin: The protein present in red blood cells responsible for carrying oxygen around the blood and releasing it where needed.

Harmonics: Higher **frequency** components of a sound that are simple ratios of the **fundamental frequency.**

Harp trap: A safe and convenient device for catching flying bats. The bats are intercepted by two parallel arrays of vertical, fine nylon threads, supported on a large metal frame. The bats fall into a soft catching bag, designed to prevent escape.

Hawking: The catching of flying prey, in flight.

Heterodyne: A simple form of bat detector that makes bat **echolocation** sounds audible to the human ear.

Heterothermy: The ability to regulate body temperature precisely over a range of temperatures.

Hibernaculum: A place where hibernating bats are found.

Hibernation: The extended use of torpor by which a bat survives the winter months when few, if any, insects are available.

Homeothermy: The maintenance of a steady, elevated body temperature by **metabolic** means. Most homeotherms function at 36-43 ºC.

Humerus: The long bone between shoulder and elbow.

Information transfer: The intentional or unintentional exchange of useful information between animals, e.g. the location of food or roosting sites.

Lactation: The production of milk and the suckling of young.

Larynx: A hollow chamber, the voice box, near the top of the windpipe or trachea. Used in the production of sound.

Lek: Display by assemblies of males to earn the right to mate with females.

Lift: The force, at right angles to the air flow over the wing, that counteracts gravity and keeps a bat in the air.

Maximum range speed: The speed at which a bat flies to travel the maximum distance on a given amount of fuel.

Megachiroptera (megabats): The Old World fruit bats or flying foxes.

Mesolithic: The age of stone tools and weapons that began at least 2.5 million years ago in Africa. It began about one million years ago in Europe, ending after 4,000 BC.

Metabolic heat: Heat generated by the body as it breaks down the chemicals derived from food. Usually a by-product of the bodies vital chemical reactions, but sometimes heat is produced for its own sake by specific tissues and reactions.

Microchiroptera (microbats): Largely insectivorous, small bats distributed around the world: all those not belonging to the **megachiroptera**!

Microclimate: Localised climate within a roost or a patch of habitat, e.g. in the shelter of trees.

Minimum power speed: The flight speed at which cost per unit time is at its lowest.

Mist net: Net of very fine nylon used to catch flying bats: requires skill and care in use.

Monoestrus: Having a single breeding season each year.

Monogamy: A single male – single female mating partnership that lasts for at least one breeding cycle.

Myoglobin: Oxygen carrying protein found in muscle.

Narrowband: A sound made up of a narrow range of **frequencies.**

Natural selection: Individuals of a species compete for resources and face the challenges posed by the environment. Those best adapted to the current conditions are those most likely to breed successfully, passing on their genes to the next generation. This process of natural selection leads to a population in which more of the individuals are better adapted to their environment, whether it is stable or in the process of changing.

Neolithic: The New Stone Age began about 8,000 years ago in the Middle East, later in Europe.

Neotropics: New World tropics: Central and South America.

Noseleaf: The often complex projection above the nostrils on many bats.

Optimal foraging: An ideal foraging strategy, for example one that maximises energy intake whilst minimising flight costs during foraging. The ideal strategy may vary depending upon prey availability, environmental conditions and the bat's goals.

Ossicles: Three small bones, the malleus, incus and stapes, in the middle ear of mammals that amplify sound.

Palaeotropics: Old World tropics of Africa, Asia and Australia.

Passive hearing: Simple hearing using sounds generated in the environment: to differentiate it from echolocation.

Phantom target or echo: An echo or 'target' generated electronically in the laboratory used to study echolocation.

Philopatry: Faithfulness to the nursery roost in which an animal was born. Most female bats of temperate species return to their natal roost to rear young throughout their lives.

Pinna: The external, fleshy part of the ear.

Pleistocene: It began 1.6 million years ago and ended about 10,000 years ago: the Ice Age.

Polygyny: When a male mates with more than one female in a single season.

Polyoestrous: Having more than one breeding cycle in a single year.

Post-calcarial lobe: A small skin flap that projects beyond the **calcar** in some bat species.

Radiotelemetry (radio-tracking): The tracking of animals by the temporary attachment of radio transmitters to study their behaviour, movements and patterns of habitat use.

Radius: One of the two bones of the forearm. The other, the ulna, has regressed in bats.

Raptor: Bird of prey.

Respiratory system: The system that takes in air and provides a very large surface area over which the oxygen in the air can diffuse into the blood: from the trachea (windpipe) down to the lungs.

Resource defence: A strategy that involves the active defence of some valuable resource, such as food, roost site or mates, from competitors.

Riparian: Habitats with freshwater.

Roost: A physical site used by bats when not flying, such as a nursery, mating, or even temporary night roost. Sometimes refers to the bats themselves in a roost.

SAC: Special Area for Conservation: one of a number of formal designations designed to protect and enhance a habitat/site.

Selective pressure: Environmental (in the broadest sense) forces that determine the direction of the adaptive responses of **natural selection.**

Sella: A part of the noseleaf of horseshoe bats.

Social call: Any call emitted by a bat that is not an echolocation call used specifically for orientation. Frequently used because we do not understand their function!

Sonar: Acronym for **SO**und **NA**vigation and **R**anging: another name for echolocation, more usually applied to technological use of the principle.

Sonogram: A graphical representation of a sound conveying information about **frequency** and intensity against time.

Spectral: For example spectral composition, the frequency components of a sound.

Spectrogram: A graphical representation of a sound: intensity plotted against frequency.

SSSI: Site of Special Scientific Interest: one of a number of formal designations designed to protect a site.

Stalling: The loss of **lift** due to the break-up of smooth airflow over the wing, usually due to an excessive **angle of attack**.

Swarming: The gathering of large aggregations of flying bats around the entrances to underground sites in late summer and autumn. Also refers to similar behaviour observed as bats return to their summer roosts at dawn, where the number of individuals involved is generally smaller and the event shorter.

Target discrimination: A measure of how easily a bat detects small insects.

Thermoregulation: The maintenance of a precise body temperature by physiological means.

Thorax: The chest

Thrust: The component of the aerodynamic force that moves a flying bat forward.

Time-expansion: The slowing down of a sound whilst preserving all its complexity, making it audible to the human ear and recordable by conventional recording systems.

Torpor: The physiological process by which bats reduce and regulate their body temperatures to save energy.

Tragus: The cartilaginous projection inside the **pinna** (external ear) of many bats.

Transect: A route taken through a habitat or series of habitats to quantitatively record bat activity.

Trawling: Capturing insects of the surface of the water using feet or tail.

Turbulence: Disturbed, chaotic flow around a wing: usually leads to loss of **lift** and **stalling**.

Ultrasound: High frequency sound beyond the human hearing range, that is greater than about 20 kHz.

Wing loading: The ratio of body weight: wing area. It has a significant influence on flight performance and efficiency.

Wing morphology: A broad term to covering wing shape and size as determinants of flight performance.

Bibliography and Other Resources

This book draws on the published and sometimes unpublished work of many dedicated biologists and conservationists. Without their efforts, very little of this book could have been written. However, to make the book accessible to as wide an audience as possible I have abandoned the usual scientific convention of attributing facts, theories and opinions to these people through their published studies. This inevitably fills the text with references that break the flow and often the train of thought. For those who wish to pursue particular topics further a bibliography is provided for each chapter. With so many sources of varying aim, scientific substance and accessibility (in terms of both getting a copy and understanding it!) deciding what to put in becomes as much a personal as a scientific decision. This is a selection of those sources that I find to be particularly informative, comprehensive or stimulating. It is intended only to give some starting points into the different topics. The emphasis is on British and European bats where this is possible or appropriate.

General and worldwide

Altringham, J.D. (1996). *Bats: Biology and Behaviour*. Oxford University Press, Oxford.

Neuweiler, G. (2000). *The Biology of Bats*. Oxford University Press, Oxford.

Both of these books are aimed primarily at the academic market, but are accessible to the keen and well-informed amateur. Both cover all the world's bats and virtually all aspects of bat biology, but Altringham gives greater emphasis to ecology and behaviour, Neuweiler to physiology. They are particularly relevant to Chapters 2–4.

Fenton, M.B. (1992). *Bats*. Facts on File, Oxford.

A large format book with many photographs and a readable, wide-ranging text.

Hill, J.E. & Smith, J.D. (1984). *Bats: a natural history*. British Museum (Natural History), London.

Written almost 20 years ago, but still full of useful information on most aspects of bat biology and natural history.

Richardson, P.W. (2000). *Bats*. Whittet Books. London.

An informative and entertaining romp through the world of bats, with cartoons and endless bad puns. Anyone can enjoy it, and it is a great present for children.

Wilson, D.E. (1997). *Bats in question – the Smithsonian answer book*. Smithsonian Institution Press, Washington.

Glossy, question-answer format book that covers all of the major issues surrounding bats and their conservation.

European

Corbet, G.B. & Harris, S., (Eds) (1991). *The handbook of British mammals*. Third edition. Blackwell Scientific, Oxford.

Chapters on all British bats, covering many aspects of their biology in a standardised and compact format. A useful reference.

Ransome, R.D. (1990). *The natural history of hibernating bats.* Christopher Helm, London.
Chapters on many aspects of the biology and natural history of British bats, with some emphasis on the author's own detailed and long-term work on greater horseshoe bats.

Schober, W. & Grimmberger, E. (1989). *Guide to the bats of Britain and Europe.* Hamlyn, London.
Sadly, out of print, this is a well-illustrated general guide to Europe's bats. Includes an identification key and other identification sections.

Mitchell-Jones, A.J., Amori, G., Bogdanowicz, W., Kryštufek, B., Reijnders, P.J.H., Spitzenberger, F., Stubbe, M., Thissen, J.B.M., Vohralík, V. & Zima, K. (1999). *The Atlas of European Mammals.* T. & A.D. Poyser, London.
Distribution maps and a short description of each species, with habitat preferences and population.

Swift, S.M. (1998). *Long-eared bats.* T. & A.D. Poyser, London.
The biology of the long-eared bats, primarily our common brown long-eared bat. Detailed account, fully referenced.

Identification – in the hand

Stebbings, R.E. (1986). *Which bat is it?* The Mammal Society and The Vincent Wildlife Trust.
Yalden, D.W. (1985). *The identification of British bats.* The Mammal Society. London.
These slim, pocket-sized booklets are really all you need to identify British bats in the hand. There is some overlap, but it is well worth having both and they are very cheap.

Greenaway F. & Hutson, A.M. (1990). *A field guide to British bats.* Bruce Coleman, London.

A pocket photographic guide to British bats with a wealth of pictures. Available from the Bat Conservation Trust.

Identification – by echolocation call

Barataud, M. (1996). *Ballades dans l'inaudible – the inaudible world.* Sitelle. Double CD and booklet.
An almost comprehensive set of recordings of European bats. Particularly useful are the time-expanded recordings, since they can be visualised using appropriate computer software.

Briggs, B. & King, D. (1998). *The bat detective.*
CD and booklet on the identification of British bats by their calls. Almost exclusively heterodyne recordings.

Russ, J. (1999). *The bats of Britain and Ireland: echolocation calls, sound analysis and species identification.* Alana Press.
How to identify bats from their echolocation calls.

Conservation

Mitchell-Jones, A.J. & McLeish, A.P., (Eds) (1999). *The bat workers' manual.* Second edition. JNCC, Peterborough.
Indispensable for the serious, practical bat worker. Chapters on everything from the law, through field techniques and safety issues, to public relations and how to grille a cave.

Practical techniques

Kunz, T.H., (Ed.) (1988). *Ecological and behavioural methods for the study of bats.* Smithsonian Institution Press, Washington.
Comprehensive descriptions of many of the techniques used to study bats. Despite recent technical advances, this is still a very useful book.

Chapter by chapter – some key scientific publications. A very limited selection of scientific publications that I believe will give a flavour of the scientific research behind the text – and a route into the scientific literature.

Chapter 2: Bats, an evolutionary success story

Altringham, J.D. (1996). *Bats: Biology and Behaviour.* Oxford University Press, Oxford.
Covers all of the topics of this chapter in greater depth. Dare I say essential reading!

Fenton, M.B. (1992). *Bats.* Facts on File, Oxford.
A large format book with many photographs and a very readable, wide-ranging text.

Hill, J.E. & Smith, J.D. (1984). *Bats: a natural history.* British Museum (Natural History), London.
Written almost 20 years ago, but still full of useful information on most aspects of bat biology and natural history.

Kunz, T.H., (Ed.) (1982). *The ecology of bats.* Plenum Press, New York.
An academic but accessible book on most aspects of bat ecology. Out of print and hard to find.

Kunz, T.H. & Fenton, M.B., (Eds) (2003). *The ecology of bats.* University of Chicago Press, Chicago.
A new book with a similar format to the last and a must buy for the serious bat biologist.

Nowak, R.M. (1994). *Walker's bats of the world.* The Johns Hopkins University Press, London.
An illustrated, systematic catalogue and description of the world's bats. A reference book, not a bedtime read, but a mine of information and an insight into the enormity of the world's bat fauna.

Schaal, S. & Zeigler, W., (Eds) (1992). *Messel: an insight into the history of life and of the earth.* Clarendon Press, Oxford.
A stunningly illustrated book on the incomparable finds in the Messel shale beds, including the first known European bat community.

Chapter 3: The biology of temperate bats

Flight

Norberg, U.M. (1990). *Vertebrate Flight.* Springer-Verlag, Berlin.
The most detailed and comprehensive book on animal flight. Although there are lots of equations, the text is often easy to follow and informative to the serious amateur.

Pennycuick, C.J. (1972). *Animal Flight.* Edward Arnold, London.
Short and now rather old, but a good, straightforward introduction to animal flight.

Tennekes, H. (1997). *The simple science of flight: from insects to jumbo jets.* MIT Press, London.
The author, an aerospace engineer, with an infectious enthusiasm and some simple maths, gives us a delightful and fascinating introduction to flight.

Echolocation

Griffin, D.R. (1958). *Listening in the dark.* Yale University Press, New Haven, Conn. Reprinted 1986.
A fascinating account of the pioneering studies into echolocation carried

out by the author. Over 60 years after his initial studies, Don Griffin is still studying bats.

Neuweiler, G. (1990). Auditory adaptations for prey capture in echolocating bats. *Physiological Reviews* **70**, 615–641.
An informative review of the literature on bat auditory adaptations up to 1990.

Suga, N. (1990). Biosonar and neural computation in bats. *Scientific American* **262**, 34–41.
An accessible account of some of the experiments and theories surrounding bat echolocation.

von der Emde, G. & Schnitzler, H.-U. (1990). Classification of insects by echolocating greater horseshoe bats. *Journal of Comparative Physiology* A **167**, 423–430.
A paper describing some very clever experiments that reveal just how much information a bat can derive from the echoes of its sonar calls.

Torpor and hibernation

Park, K.J., Jones, G. & Ransome, R.D. (2000). Torpor, arousal and activity of hibernating greater horseshoe bats (*Rhinolophus ferrumequinum*). *Functional Ecology* **14**, 580–588.
Speakman, J.R. & Racey, P.A. (1989). Hibernal ecology of the pipistrelle bat: energy expenditure, water requirements and mass loss, implications for survival and the function of winter emergence flights. *Journal of Animal Ecology* **58**, 797–813.
Thomas, D.W. (1995). *The physiological ecology of hibernation in vespertilionid bats*. In: **Racey, P.A. & Swift, S.W.,** (Eds) *Ecology, evolution and behaviour of bats*. Oxford

University Press, Oxford. Pp. 233–244.
A selection of papers using a wide range of approaches that gives an up to date insight into research on hibernation.

Life history cycles and reproduction

Kunz, T.H. & Hood, W.R. (2000). Parental care and postnatal growth in the Chiroptera. In: **Crichton, E.G. & Krutczsch, P.H.,** (Eds) *Reproductive biology of bats*. Academic Press, London. Pp 415–468.
McCracken, G.F. & Wilkinson, G.S. (2000). Bat mating systems. In: **Crichton, E.G. & Krutczsch, P.H.,** (Eds) *Reproductive biology of bats.* Academic Press, London. Pp 321–362.
Racey, P.A. & Entwistle, A.C. (2000). Life-history and reproductive strategies of bats. In: **Crichton, E.G. & Krutczsch, P.H.,** (Eds) *Reproductive biology of bats*. Academic Press, London. Pp. 363–414.
Three chapters of a recent book that cover many of the general concepts and much of the detailed biology of bat reproduction, social systems and life history patterns. The perspective is global, but temperate bats are covered and British bats when possible.

Chapter 4: An ecological synthesis

Norberg, U.M. & Rayner, J.M.V. (1987). Ecological morphology and flight in bats (Mammalia; Chiroptera): wing adaptations, flight performance, foraging strategy and echolocation. *Philosophical Transactions of the Royal Society of London B* **316**, 335–427.
A detailed and comprehensive analysis of the links between flight and ecology.

Jones, G. & Rayner, J.M.V. (1988). Flight performance, foraging tactics and echolocation in free-living Daubenton's bats, *Myotis daubentonii. Journal of Zoology* **215**, 113–132.

Jones, G. & Rayner, J.M.V. (1989). Foraging behaviour and echolocation of wild horseshoe bats *Rhinolophus ferrumequinum* and *R. hipposideros. Behavioural Ecology and Sociobiology* **25**, 183–191.

Siemers, B.M. & Schnitzler, H.-U. (2000). Natterer's bat (*Myotis nattereri*) hawks for prey close to vegetation using echolocation calls of very broad bandwidth. *Behavioural Ecology and Sociobiology* **47**, 400–412.

Some of the papers describing studies of foraging strategies of British bats in the wild.

Jones, G. (1994). Scaling of wingbeat and echolocation pulse emission rates in bats: why are aerial insectivores so small? *Functional Ecology* **8**, 450–457.

One reason why bats are small.

Fenton, M.B. (1990). The foraging behaviour of animal-eating bats. *Canadian Journal of Zoology* **68**, 411–422.

A review of foraging behaviour in echolocating, primarily insectivorous, bats.

Jones, G. (1990). Prey selection by the greater horseshoe bat (*Rhinolophus ferrumequinum*): optimal foraging by echolocation. *Journal of Animal Ecology* **59**, 587–602.

A detailed study of seasonal changes in diet as evidence for selective, optimal foraging.

Kerth, G., Weissmann, K. & König, B. (2001). Day roost selection in female Bechstein's bats (*Myotis bechsteinii*): a field experiment to determine the influence of roost temperature. *Oecologia* **126**, 1–9.

A recent paper investigating the importance of roost temperature to roost switching in Bechstein's bats.

Entwistle, A.C., Racey, P.A. & Speakman, J.R. (1997). Roost selection by the brown long-eared bat *Plecotus auritus. Journal of Applied Ecology* **34**, 399–408.

Jenkins, E.V., Laine, T., Morgan, S.E., Cole, K.R. & Speakman, K.R. (1998). Roost selection in the pipistrelle bat, *Pipistrellus pipistrellus*, in northeast Scotland. *Animal Behaviour* **56**, 909–917.

Papers investigating what makes a good roost.

Entwistle, A.C., Racey, P.A. & Speakman, J.R. (1996). Habitat exploitation by a gleaning bat, *Plecotus auritus. Philosophical Transactions of the Royal Society of London* B **351**, 921–931.

Waters, D.A., Jones, G. & Furlong, M. (1999). Foraging ecology of Leisler's bat (*Nyctalus leisleri*) at two sites in southern England. *Journal of Zoology* **249**, 173–180.

Schofield, H. & Morris, C. (2000). Ranging behaviour and habitat preferences of female Bechstein's bat, *Myotis bechsteinii*, in summer. *Report of Vincent Wildlife Trust, Ledbury*.

Recent papers investigating habitat preferences and foraging behaviour of British bats.

Burland, T.M., Barratt, E.M., Nichols, R.A. & Racey, P.A. (2001). Mating patterns, relatedness and the basis of natal philopatry in the brown long-eared bat, *Plecotus auritus. Molecular Ecology* **10**, 1309–1321.

Kerth, G., Wagner, M & König, B. (2001). Roosting together, foraging apart: information transfer about food is unlikely to explain sociality in female Bechstein's bats (*Myotis bechsteinii*). *Behavioural Ecology and Sociobiology* **50**, 283–291.

Petit, E. & Mayer, F. (1999). Male dispersal in the noctule bat (*Nyctalus noctula*): where are the limits? *Proceedings of the Royal Society of London* B **266**, 1717–1722.

Rossiter, S.J., Jones, G., Ransome, R.D. & Barratt, E.M. (2000). Parentage, reproductive success and breeding behaviour in the greater horseshoe bat (*Rhinolophus ferrumequinum*). *Proceedings of the Royal Society of London B* **267**, 545–551.

Some recent papers illustrating how molecular genetics is being used to help unravel complex behavioural and ecological questions about bats.

Vaughan, N. (1997). The diets of British bats (Chiroptera). *Mammal Review* **27**, 77–94.

A detailed review of what British bats feed on.

Jones, G. & Rydell, J. (2002). In: **Kunz, T.H. & Fenton, M.B.,** (Eds) (2003). *The ecology of bats.* University of Chicago Press. Chicago.

Review of the arms race between insectivorous bats and their prey.

Speakman, J.R. (1991). The impact of predation by birds on bat populations in the British Isles. *Mammal Review* **21**, 132–142.

An assessment of how predatory birds may impact bat populations. Includes estimates of UK bat populations.

Jones, G. & Rydell, J. (1994). Foraging strategy and predation risk as factors influencing emergence time in echolocating bats. *Philosophical Transactions of the Royal Society of London* B **346**, 445–455.

Speakman, J.R., Stone, R.E. & Kerslake, J.E. (1995). Temporal patterns in the emergence behaviour of pipistrelle bats, *Pipistrellus pipistrellus*, from maternity colonies are consistent with an anti-predator response. *Animal Behaviour* **50**, 1147–1156.

Two different aspects of how predation risk can influence bat behaviour.

Chapter 5: British bats, past and present

Schaal, S. & Zeigler, W., (Eds) (1992). *Messel: an insight into the history of life and of the earth.* Clarendon Press. Oxford.

A stunningly illustrated book on the incomparable finds in the Messel shale beds, including the first known European bat community.

Yalden, D.W. (1999). *The history of British mammals.* T. & A.D. Poyser. London.

A comprehensive account of British mammalian history that covers the sparse literature on bats.

Mitchell-Jones, A.J., Amori, G., Bogdanowicz, W., Kryštufek, B., Reijnders, P.J.H., Spitzenberger, F., Stubbe, M., Thissen, J.B.M., Vohralík, V. & Zima, K. (1999). *The Atlas of European Mammals.* T. & A. D. Poyser, London.

Distribution maps and a short description of each species, with habitat preferences and population.

Corbet, G.B. & Harris, S., (Eds) (1991). *The handbook of British mammals.* Third edition. Blackwell

Scientific. Oxford.
Chapters on all British bats, covering many aspects of their biology in a standardised and compact format. A useful reference.

Richardson, P.W. (2000). *Distribution atlas of bats in Britain and Ireland, 1980–1999.* Bat Conservation Trust. London.
The most recent summary of bat distribution in Britain, includes maps and a little explanatory text.

Hutson, A.M. (1993). *Action plan for the conservation of bats in the United Kingdom.* Bat Conservation Trust, London.
Includes estimates of UK bat populations.

Jones, G. & van Parijs, S.M. (1993). Bimodal echolocation in pipistrelle bats: are cryptis species present? *Proceedings of the Royal Society of London B* **251**, 119–125.
Park, K.J., Altringham, J.D. & Jones, G. (1996). Assortative roosting in the two phonic types of *Pipistrellus pipistrellus* during the mating season. *Proceedings of the Royal Society of London B* **263**, 1495–1499.
Barlow, K.E. & Jones, G. (1997). Differences in songflight calls and social calls between two phonic types of the vespertilionid bat *Pipistrellus pipistrellus.* *Journal of Zoology* **241**, 315–324.
Barratt, E.M., Deaville, R., Burland, T.M., Bruford, M.W., Jones, G., Racey, P.A. & Wayne, R.K. (1997). DNA answers the call of pipistrelle bat species. *Nature* **387**, 138–139.
Some of the papers behind the discovery that the common pipistrelle was two distinct species.

Mayer, F. & von Helverson, O.

(2001). Cryptic diversity in European bats. *Proceedings of the Royal Society* B **268**, 1825–1832.
A genetic analysis of European bats, raising the possibility of more 'hidden' species and questioning our ability to define species on the basis of either morphology or genetics alone.

Chapter 6: Conservation

Entwistle, A.C. et al. (2001). *Habitat management for bats.* JNCC, Peterborough.
A practical guide structured around habitats and individual species.

www.batcon.org
Website for Bat Conservation International. An extensive and attractive website, rich in resources and links to other useful websites.

www.bats.org.uk
Website for the Bat Conservation Trust. Basic information on bats and bat conservation in the UK, with links to other sites.

Other useful websites:
www.abdn.ac.uk/mammal
The Mammal Society
www.mammalstrustuk.org
Mammals Trust UK
www.ptes.org
Peoples Trust for Endangered Species
www.vwt.org.uk
Vincent Wildlife Trust
www.fauna-flora.org
Fauna and Flora International
www.nationaltrust.org.uk
The National Trust
www.english-nature.org.uk
English Nature
www.ccw.gov.uk
Countryside Council for Wales
www.snh.org.uk
Scottish Natural Heritage
www.eurobats.org

Eurobats Secretariat
www.forestry.gov.uk
Forestry Commission

Chapter 7: Watching and studying British bats

Many of the papers listed under Chapters 4 and 5 will be relevant here too.

Mitchell-Jones, A.J. & McLeish, A.P., (Eds) (1999). *The bat workers' manual.* Second edition. JNCC. Peterborough.
Indispensable for the serious, practical bat worker. Chapters on everything from the law, through practical techniques and safety issues, to public relations.

Vaughan, N., Jones, G. & Harris, S. (1997). Identification of British bat species by multivariate analysis of echolocation call parameters. *Bioacoustics* **7**, 189–207.
The first detailed paper on analysis of time-expanded bat calls. You may not want to attempt the multivariate analysis, but the summary table and figures are invaluable.

Kunz, T.H., (Ed.) (1988). *Ecological and behavioural methods for the study of bats.* Smithsonian Institution Press. Washington.
Comprehensive descriptions of many of the techniques used to study bats.

Park, K.J., Masters, E. & Altringham, J.D. (1998). Social structure of three sympatric bat species (Vespertilionidae). *Journal of Zoology* **244**, 379–389.
An analysis of data collected from a successful bat box scheme in Dorset.

Walsh, A.L. & Harris, S. (1996). Determinants of vespertilionid bat abundance in Britain: geographical, land class and local

habitat relationships. *Journal of Applied Ecology* **33**, 519–529.
Habitat preference study on a large scale.

Vaughan, N., Jones, G. & Harris, S. (1997). Habitat use by bats (Chiroptera) assessed by means of a broad-band acoustic method. *Journal of Applied Ecology* **34**, 716–730.
Habitat preference on a smaller scale.

Warren, R.D., Waters, D.A., Altringham, J.D. & Bullock, D.J. (2000). The distribution of Daubenton's bats *Myotis daubentonii* and pipistrelles *Pipistrellus pipistrellus* in relation to small-scale features of riparian habitat. *Biological Conservation* **92**, 85–91.
Habitat preference study on a small scale: the importance of local landscape features.

Barnard, C., Gilbert, F. & McGregor, P. (2001). *Asking questions in biology.* Second edition. Prentice Hall, London.
Fowler, J., Cohen, L. & Jarvis, P. (1998). *Practical statistics for field biology. Second edition.* John Wiley and Sons, Chichester.
Two books to help you design your experiments and make sense of the results.

Chapter 8: Practical conservation projects

See resources for Chapter 6.

Chapter 9: Identification: how to identify a bat in the hand

Stebbings, R.E. (1986). *Which bat is it?* The Mammal Society and The Vincent Wildlife Trust.
Yalden, D.W. (1985). *The identification of British bats.* The Mammal Society. London.

These slim, pocket-sized booklets are really all you need to identify British bats in the hand. There is some overlap, but it is well worth having both and they are very cheap.

Greenaway F. & Hutson, A.M. (1990). *A field guide to British bats.* Bruce Coleman, London.

A pocket photographic guide to British bats with a wealth of pictures. Available from the Bat Conservation Trust.

Resources: websites for equipment and software

This is not comprehensive, but lists those I consider to be most useful.

Supplier of a wide range of equipment for bat study

www.alanaecology.com
  Alana Ecology

Bat detectors

www.magenta2000.co.uk
  Magenta Electronics
www.bahnhof.se/~pettersson/
  Pettersson Electronik

www.skyeinstruments.com
  Skye Instruments
www.batbox.com
  Stag Electronics
www.users.globalnet.co.uk/~court-pan
  Tranquility
www.ultrasoundadvice
  Ultrasound Advice

Sound analysis software

www.bahnhof.se/~pettersson/
  Pettersson Electronik
www.visualizationsoftware.com
  Spectrogram
www.syntrillium.com
  Syntrillium

Data loggers

www.geminidataloggers.com
  Gemini Dataloggers

Websites and other sources of information

www.batcon.org
  Bat Conservation International
www.bats.org.uk
  Bat Conservation Trust

Index